AF294779

BG Chemie

Springer

Berlin
Heidelberg
New York
Barcelona
Budapest
Hong Kong
London
Milan
Paris
Santa Clara
Singapore
Tokyo

Toxicological Evaluations

11

Potential Health Hazards
of Existing Chemicals

 Springer

Berufsgenossenschaft
der chemischen Industrie
Kurfürsten-Anlage 62

D 69115 Heidelberg

Library of Congress Cataloging-in-Publication Data (Revised for vol. 11)
Toxicological evaluations.
Includes bibliographical references and indexes.
1. Toxicity testing. I. Berufsgenossenschaft der Chemischen Industrie. [DNLM: 1 Hazardous
Substances Toxicity. WA 455 T7549
RA1199.T678; 1990; 615.9'07; 90-10020
ISBN-13:978-3-642-64423-8   e-ISBN-13: 978-3-642-60477-5
DOI: 10.1007/978-3-642-60477-5

ISBN-13:978-3-642-64423-8 Springer-Verlag Berlin Heidelberg New York

© Springer-Verlag Berlin Heidelberg 1997
ISSN X-8038-3000-7
Softcover reprint of the hardcover 1st edition 1997

Typesetting: Medio GmbH, Berlin

SPIN 10537106   51/3020 - 5 4 3 2 1 0 - Printed on acid-free paper

Preface

As part of its "Programme for the prevention of health hazards caused by industrial substances", the Berufsgenossenschaft der chemischen Industrie (BG Chemie, Employment Accident Insurance Fund of the Chemical Industry) began in 1977 to investigate the toxicity of those chemicals which are widely used, have many different applications and are suspected of being dangerous to health, in particular of having long-term effects. The investigations consist of a literature search and – depending on the results – commissions of experimental studies. It is hoped by means of this testing to close gaps in our knowledge and to increase the scientific validity of the required risk assessments. The results of the toxicological investigations carried out by BG Chemie, and the resulting substance assessments have been published in German since 1987 in the form of 187 "Toxikologische Bewertungen" ("Toxicological Evaluations") up to now.

In order to make this useful information internationally available, BG Chemie began in October 1990 to publish them as a book series in English, of which the eleventh volume (containing 11 individual evaluations) is presented here. Therefore for 150 existing chemicals, "Toxicological Evaluations" are available in English at the moment, a further 58 are in preparation and will be published soon.

Because of the short time between publishing volume 4 - 11, printing of an "Introduction" (consisting of a general overview of the programme, lists with names of people involved as well as substances under investigation) has been abandoned in this volume and will be included in volume 12 again. If more detailed information is required about the ongoing work, see volume 1 or 4 or contact BG Chemie at first hand.

BG Chemie hopes that, for many people working in the chemical industry, this information will be of practical help in assessing hazards to health at the workplace.

Dr. Radek

Contents

Anthraquinone

1. Summary and assessment

Following intravenous injection of ^{14}C-labelled anthraquinone, the radioactivity is rapidly distributed through the tissues of rats. Measurable elimination only begins 8 hours after injection. In rats, absorption begins only a few minutes after oral administration of radiolabelled anthraquinone, and is almost total. In relation to the total radioactivity, maximum plasma levels following oral administration of 1 mg anthraquinone/kg body weight are measured in male rats after 5 hours and in female rats after 12 hours. In the rat, about one third of the administered anthraquinone and its metabolites are excreted in the urine and two-thirds in the faeces, irrespective of the route of administration or the dose given. Only a very small proportion of the administered radioactivity (<0.01%) is excreted in the breath. The half-life for elimination of total radioactivity following intravenous or oral administration of 1 mg ^{14}C-labelled anthraquinone/kg body weight is ca. 20 hours. The majority of the radioactivity is excreted within one day. Only traces of radioactivity (1.5 to 5%) are still detectable in the organs and tissues of rats 96 hours after administration of the substance, and it does not accumulate in any specific organ or tissues. Anthraquinone is subjected to marked enterohepatic circulation in rats following enteral administration. The intensive excretion of radioactivity in the faeces following intravenous injection of ^{14}C-labelled anthraquinone also demonstrates the importance of biliary elimination in this species. This has been confirmed in studies in rats with biliary fistulae, which excreted 35% of the administered radioactivity in the bile following intravenous injection. The main faecal excretion product in rats is unchanged anthraquinone. In addition to this, ca. 4% free 2-hydroxyanthraquinone is found in the faeces. The main metabolite in the urine is conjugated

2-hydroxyanthraquinone (about 20% of the total radioactivity). In addition 1-hydroxyanthraquinone, free 2-hydroxyanthraquinone, 1% unchanged anthraquinone and other, as yet unidentified, metabolites are excreted in the urine. Unchanged anthraquinone is also detectable analytically in the faeces of mice after oral administration.

Technical grade anthraquinone is of very low toxicity in mice and rats on oral and intraperitoneal administration with LD_{50} values of >5.0 g/kg body weight (observation period 14 days). Rats have also survived higher doses of anthraquinone (10, 15 and 20 g/kg body weight). Compared with small rodents, sheep are more sensitive to anthraquinone as, on oral administration of graduated doses, dose-dependent reductions in feed intake and body weight occur, as well as late deaths. Doses of between 0.15 and 0.30 g anthraquinone/kg body weight are lethal in sheep following a 14-day observation period. Anthraquinone does not lead to morphological effects on the internal organs and tissues of mice, rats or sheep other than non-specific irritant effects following intraperitoneal injection. Anthraquinone is also of low toxicity on dermal application to rats (LD_{50} >5.0 g/kg body weight, observation period 14 days) and rabbits (LD_{50} >3.0 g/kg body weight, scarified skin, observation period 14 days) and causes no morphological effects on the internal organs and tissues of the animals. Inhalation exposure for 4 hours to technical grade anthraquinone at the maximum concentration that could be produced with a dust generator (measured at the nose level of the rat as 1327 mg/m^3) was tolerated by rats without signs of toxicity, and did not cause any treatment-related effects on the organs and tissues. Technical grade anthraquinone can also be regarded as being of low acute toxicity with a 4-hour LC_{50} of >1327 mg/m^3 (observation period 14 days).

Repeated oral administration of anthraquinone in daily doses of 2 mg/kg body weight for 28 days is tolerated by rats without adverse effects. Higher doses of anthraquinone (10, 20, 50 and 250 mg/kg body weight/day) lead to reduced body weight gain, indications of anaemia, and increases in organ weights, particularly the relative weights of the spleen and liver. The effect on spleen weight is explained by congestion, that on liver weight by cell enlargement, probably as a consequence of an adaptation of the liver due to stimulation of the enzymes that degrade foreign compounds.

Pure and technical grade anthraquinone, and a 25% anthraquinone preparation, are not irritating to the skin or the eyes of the rabbit.

Anthraquinone does not induce skin sensitisation in guinea pigs and is not phototoxic in hairless mice after dermal or intraperitoneal application.

In a 3-month feeding study in rats, an anthraquinone concentration of 15 ppm did not cause damage, but cannot be regarded as a no effect level as the body weight gain of the animals was reduced because of their lowered feed intake during the first weeks of the study. The increased liver weight and liver cell enlargement, particularly in the highest dose group (150 ppm), indicate an adaptation of the liver through stimulation of enzymes that degrade foreign compounds, as in the 28-day rat study. The no effect level for exposure of rats to anthraquinone as dust in a dynamic system for 4 months is 5.2 mg/m^3. An anthraquinone dust concentration of 12.2 mg/m^3 led to a slight decrease in body weight gain and indications of anaemia.

Pure anthraquinone did not induce mutations in the Salmonella/microsome test in strains TA 1535, TA 100, TA 1537, TA 98 or TA 1538 either with or without S9-mix as a metabolic activation system. It was also not genotoxic in a TA 1535 strain with plasmid pSK 1002 (the so-called umu test) or in the 8-azaguanine-resistant strain TM 677, with or without the addition of S9-mix as a metabolic activation system. However, technical grade anthraquinone can cause mutations in the Salmonella/microsome test due to contaminants.

An increased number of low molecular weight fragments was found in the nuclei of tissues from the liver and kidney of anthraquinone-treated mice, compared with nuclei from liver and kidney tissues from control mice. This is indicative of the formation of DNA strand breaks. The effect of anthraquinone under these study conditions is less pronounced than that of known carcinogens such as benzidine, 4-aminotoluene or N-nitroso-N-methylurea. DNA strand breaks can also be caused through non-specific toxic effects.

Anthraquinone is not mutagenic in the micronucleus test in the mouse after intraperitoneal injection.

Anthraquinone was not carcinogenic in 4 preliminary carcinogenicity studies in mice involving repeated oral, subcutaneous and dermal exposure. Despite the limited strength of evidence from these 4 early studies, it can be assumed that pure anthraquinone has no carcinogenic potential because it is not genotoxic and is negative in the cell transformation test.

Anthraquinone does not disturb normal cell differentiation in embryonic Drosophila tissue and has thus shown no indications of teratogenicity in this test system. This study does not allow an opinion to be formed on the teratogenic potential in mammals.

When present in an equimolar dose to the carcinogen 3'-methyl-4-dimethylaminoazobenzene, anthraquinone inhibits binding of the carcinogen to rat liver protein, reduces the glutathione content in the liver and decreases the activity of deoxyribonuclease from bovine pancreas. The activities of numerous other enzymes are also affected by anthraquinone.

Following daily intraperitoneal injection of mice with a 0.1% anthraquinone solution, the growth of Twort carcinoma, a transplantable mouse tumour, is inhibited by up to 47% when compared with that in control mice. Treatment of NF mouse sarcoma fragments with anthraquinone does not inhibit the growth of the tumour after reimplantation. The cytotoxic effect of anthraquinone (0.1 mg/ml) on suspensions of human malignant ovarian tumour cells is increased by the addition of DMSO.

In vivo, anthraquinone has a sedative effect on mammals, but is not analgesic or antipyretic.

Anthraquinone is not cytotoxic to peritoneal or lung macrophages in vitro, and is not fibrogenic in rats in vivo (intraperitoneal injection of 50 mg anthraquinone/kg body weight).

No primary irritant or phototoxic effects were seen in skin tests with anthraquinone in a man who had been sporadically occupationally exposed to anthraquinone and many other chemicals. In another worker who was employed in processing cellulose pulp, subacute dermatitis (which recurred after every addition of anthraquinone) was judged to be a phototoxic reaction triggered by anthraquinone or a contaminant present in the batches of anthraquinone used (phenanthrene, anthracene, anthrone, nitrobenzene), based on photo-patch tests. High concentrations of anthraquinone in the air of workplaces (10 to 1650 mg/m^3) can cause headaches, general debility and irritation of the eyes and unprotected areas of skin. Furthermore, a reduced ability to concentrate and indications of possible impairment of the function of the CNS and of the regulation of the circulation have been described in workers employed in anthraquinone plants for several years. It is not known whether the effects on functional parameters measured in exposed subjects are caused by anthraquinone or by its major contaminant, anthracene.

90-Day oral studies in F344 rats and B6C3F1 mice carried out as part of the National Toxicology Program (NTP) are currently being evaluated.

2. Name of substance

2.1	Usual name	Anthraquinone
2.2	IUPAC-name	9,10-Dioxoanthracene
2.3	CAS-No.	84–65–1
2.4	EINECS-No.	201–549–0

3. Synonyms, common and trade names

9,10-Anthracenedione
Anthracene-9,10-quinone
Anthrachinon
9,10-Anthraquinone
Corbite
9,10-Dihydro-9,10-
   diketoanthracene
9,10-Dihydro-9,10-
   dioxanthracene
Diphenylketone
Hoelite
Morkit DS 25 (anthraquinone
   content 25%)

Morkit WP 80 (anthraquinone
content 80%)
Sibutol Morkit FS 375
(anthraquinone content 15%)

4.   Structural and molecular formulae

4.1   Structural formula

4.2   Molecular formula $C_{14}H_8O_2$

5.   Physical and chemical properties

5.1   Molecular mass, g/mol   208.20

5.2   Melting point, °C   284   (Bayer, 1993a)

5.3   Boiling point, °C   377 (at 1013 hPa) (Bayer, 1993a)

5.4   Vapour pressure, hPa   0.000013 (at 68.8 °C)
                                   1.3 (at 190 °C)   (Bayer, 1993a)

5.5   Density, g/cm³   1.44 (at 20 °C)   (Bayer, 1993a)

5.6   Solubility in water   0.125 mg/l (at 22 °C)
                                                    (Bayer, 1993a)

5.7   Solubility in   0.05 g/100 g ethanol (at 18 °C)
       organic solvents   0.44 g/100 g ethanol (at 25 °C)
                                  2.25 g/100 g in boiling ethanol
                                  0.11 g/100 g ether (at 25 °C)
                                  0.61 g/100 g chloroform
                                    (at 20 °C)
                                  1.00 g/100 g chloroform (at 40 °C)
                                  1.60 g/100 g chloroform
                                    (at 60 °C)

0.26 g/100 g benzene (at 20 °C)
0.50 g/100 g benzene (at 40 °C)
1.00 g/100 g benzene (at 60 °C)
1.80 g/100 g benzene (at 80 °C)
0.30 g/100 g toluene (at 25 °C)
(Budavari et al., 1989)

5.8   Solubility in fat   No information available

5.9   pH value   No information available

5.10   Conversion factor $1\ ml/m^3$ (ppm) $\cong 8.50\ mg/m^3$
$1\ mg/m^3 \cong 0.12\ ml/m^3$ (ppm)
(at 1013 hPa and 25 °C)

6.   Uses

As a raw material in dyestuff and pigment production; as a processing aid in textile dyeing and cellulose manufacture; as a bird-repel-
lent on cereal and vegetable crops; as a fungicide (Bayer, 1993a;
Budavari et al., 1989; Copplestone, 1983; Cofrancesco, 1991; Munn
and Smagghe, 1983; Vogel, 1985).

7.   Experimental results

7.1   Toxicokinetics and metabolism

Following administration of 100 mg anthraquinone (purity unspecified) to rats (weight 115 to 200 g, strain unspecified; dose equivalent to 500 to 750 mg anthraquinone/kg body weight), 2-hydroxy-
anthraquinone was detected in the urine (time period over which
urine collected unspecified). The amount of 2-hydroxyanthra-
quinone excreted represented only a few percent of the anthraquinone given to the rats. Detection of the substance was achieved
by determination of the melting point, solubility in organic sol-

vents and alkali and measurements of adsorption (Sato et al., 1956).

In a further study, rats were given the same amount of anthraquinone orally and then injected subcutaneously with 100 μCi ^{35}S-sulfate. The urine was then collected over 24 hours. In this study, it was shown by means of paper chromatography and colour reactions that, in addition to 2-hydroxyanthraquinone, the sulfate conjugate of 2-hydroxyanthraquinone was also eliminated in the urine of the rats (Sato et al., 1959).

4 rats (strain unspecified) were given feed containing 5% (w/w) anthraquinone (specification not provided) for 4 days. The urine of the rats was collected daily and the substances excreted in it analyzed by means of thin layer chromatography following chemical and enzymatic hydrolysis. The main products excreted were 2-hydroxy-9,10-anthraquinone, its sulfuric acid and glucuronic acid conjugates, anthrone and 9-hydroxyanthranyl sulfate (Sims, 1964).

Anthraquinone did not have a laxative effect in Swiss albino mice on oral administration of doses of up to 10 mg/20 g mouse (500 mg/kg body weight), in contrast to a number of pure glycosidic anthraquinone compounds. Each dose group contained 12 mice which were given no feed or drinking water from the previous evening up until 24 hours after administration of the test substance. The faeces of the mice, which were kept in individual cages, were collected on underlying white paper. The consistency of the faeces was determined every 30 minutes. In addition, the urine of the mice was collected and both the urine and faeces were analyzed (thin layer chromatography and HPLC). In the studies with anthraquinone, the unchanged substance was found in the faeces, but not in the urine. No anthraquinone metabolites were detected analytically in the faeces or the urine of the mice. Maximum excretion occurred 4 hours after administration of the substance, but traces of anthraquinone were still detectable in the faeces 24 hours after gavaging. The authors suggested that a polycondensated hydroxyanthrone glycoside (hypericin ?) was probably present in the extracts tested, and this had a synergistic effect with the glycosides but not with anthraquinone. A "reduction of anthraquinone ions" to anthrones was excluded as a mechanism for the reduction in the resistance of the faeces and inhibition of the sodium pump was discussed. Metabolic degradation of the aglycones occurred by hydroxylation, oxidation and demethylation. In this way, hydroxy-derivatives and glucuronides arose (Longo, 1980).

The biokinetic behaviour in rats was studied with anthraquinone labelled with ^{14}C at positions 9 and 10. The specific radioactivity was 170 μCi/mg, and the radiochemical purity ca. 100%. Male rats were treated with the substance at doses of 3.0, 1.0 and 0.1 mg/kg body weight orally, 0.1 mg/kg body weight intraduodenally and 1.0 mg/kg body weight intravenously, while female rats received 1.0 mg/kg body weight orally. Prior to administration, the chloroform stock solution that contained the labelled substance was mixed with Cremophor EL and made up to the appropriate final volume with physiological saline. The total radioactivity remaining in the urine and faeces, in the expired air and in the rats and their tissues was measured over time. The results thus related to the sum of the unchanged substance and its labelled metabolites. Following enteral administration, the radioactive anthraquinone was almost completely absorbed. Absorption began after ca. 2 to 3 minutes (see Table 1).

Table 1. Kinetic data – based on total radioactivity – following acute intravenous or oral treatment of rats (MURA:SPRA, SPF 68 Han) with [9, 10 -^{14}C]-labelled anthraquinone (Bayer, 1983b)

Dose (mg/kg bw)	Sex	Absorption half-lives (hours)		Elimination half-lives (hours)	Plasma	
		a_1	a_2		relative c_{max} (P)	t_{max} (hours)
intravenous						
1.0	male	–	–	19[1]	–	–
oral						
0.1	male	0.65	–	5.6	0.75	2.5
1.0	male	0.08	1.3	20	0.46	5.1
1.0	female	0.17	4.8	20	0.43	12.0

[1] terminal elimination half-life, calculated from the values between 8 and 48 hours after administration

[2] P: measured activitiy per g plasma divided by the activity administered per g body weight
bw body weight

Absorption began 2 minutes after oral administration of 0.1 mg/kg to male rats. The initial plasma half-life was ca. 40 minutes; the maximum plasma level was reached after ca. 2½ hours. The relative c_{max} (maximum plasma concentration) was 0.75. Following oral administration of 1 mg anthraquinone/kg, the t_{max} (time point at which the maximum plasma concentration was reached) was 5 hours in male rats and 12 hours in female rats. The radioactivity was eliminated slowly. The elimination half-life after oral administration of 1 mg/kg was 20 hours in both male and female rats (see Table 1). After intravenous injection, the radioactivity was rapidly distributed in the tissues. However, measurable elimination began only 8 hours after injection. Only a very small proportion of the administered radioactivity was excreted in the breath; less than 0.01% of the administered activity was exhaled within 48 hours of oral administration of 1 mg/kg (see Table 2).

Table 2. Excretion and distribution of total radioactivity in rats (MURA:SPRA, SPF 68 Han) 48 hours after a single intravenous, intraduodenal or oral dose of [9, 10-^{14}C]-labelled anthraquinone (Bayer, 1983b)

Dose (mg/kg bw)	Numbers of animals and sex	Amount as % of total radioactivity						Recovery rate in %
		Urine	Faeces	Body without GI tract	GI tract	Bile	CO$_2$ exhaled air	
intravenous								
1.0	10 male	36.3	56.8	3.5	0.6	–	–	106
intraduodenal								
0.1	10 male	23.7	1.6	6.3	0.4	64.5	–	108
oral								
0.1	10 male	33.0	59.8	6.0	0.7	–	–	101
1.0	18 male	37.5	56.6	3.9	0.8	–	<0.01 (8 male)	103
1.0	10 female	35.0	59.0	5.7	1.1	–	–	99

bw body weight
GI gastro-intestinal tract

It can be concluded that the labelling position in the molecule is stable as regards degradation to C_1 fragments, which can be eliminated in the breath as $^{14}CO_2$. Within 48 hours, 33 to 37.5% of the recovered radioactivity had been excreted via the kidneys and 56.6 to 59.8% in the faeces in both male and female rats in all dose groups studied, after intravenous and oral administration. The total excreted in the urine and faeces within 48 hours was on average ca. 95% of the total radioactivity. The amount excreted was proportional to the dose administered and was independent of sex and route of administration. Between 3 and 6.5% of the recovered radioactivity was found in the body (without the gastro-intestinal tract) 48 hours after administration. The highest concentrations were determined in the liver and kidneys, followed by the plasma and erythrocytes. The values for the thyroid gland, heart, spleen, adrenal glands and skin were in the range of those for the rest of the body (without the gastro-intestinal tract). The concentrations in the adipose tissue, testes and muscles were lower by a factor of about 3, while that in the brain was lower by a factor of about 7. By means of whole body autoradiography, the distribution of radioactivity was determined 5 minutes after intravenous injection (1 mg/kg) and 2, 4, 8, 24 and 48 hours after oral administration (3 mg/kg). Increased radioactivity relative to the blood was detected around the brown adipose tissue of the kidney cortex, the roots of the teeth and the Cowper's gland, 5 minutes after intravenous injection of 1 mg/kg. The concentrations of radioactivity in the adrenal cortex and the central nervous system were lower, but were still higher than that in the blood. 2 hours after oral administration of 3 mg/kg, the highest radioactivity was found within the gastro-intestinal tract, in accordance with the route of administration. The distribution pattern of the radioactivity in the other tissues and organs largely corresponded with the primary distribution. The liver could be more easily identified and higher concentrations were also evident in the white adipose tissue. The radioactivity in the animals was clearly reduced 48 hours after oral administration of 3 mg/kg. Higher levels of radioactivity relative to other tissues and organs were only found in the liver, large intestine and kidneys and in the mucous membranes of the nose. Rats with biliary fistulae excreted $^2/_3$ of the total radioactivity in the bile after intraduodenal administration of

0.1 mg/kg (see Table 2). Direct comparison of renal and biliary excretion of radioactivity in intact rats and those with biliary fistulae showed that marked enterohepatic circulation occurred, in the course of which the radioactivity was excreted proportionally via the kidneys and in the faeces. This long-lasting enterohepatic circulation was also substantiated by the high levels of radioactivity visible in the liver and gastro-intestinal tract contents by whole body autoradiography 48 hours after administration of the substance (Bayer, 1983b).

Further metabolism studies were carried out with 9,10-^{14}C-anthraquinone in Sprague-Dawley rats which were given an oral dose of 5 mg/kg body weight. The specific radioactivity was 170 µCi/mg and the radiochemical purity was greater than 99%. The test substance was administered orally in physiological saline containing 10% Cremophor EL and 0.5% traganth. The urine was collected between 0 and 8 hours and 8 and 24 hours after administration, while the faeces were collected between 0 and 24 hours. Of the radioactivity eliminated, 60% was excreted in the faeces and 40% in the urine. Of the faecal radioactivity, 75% could be extracted with methanol. The main faecal excretion product, anthraquinone, was detected by inverse dilution analysis. In addition about 4% free 2-hydroxyanthraquinone was found in the faeces. The urine contained conjugated 2-hydroxyanthraquinone as the main metabolite (about 20% of the total radioactivity). This was detected, following hydrolysis, by HPLC, GC/MS and inverse dilution analysis. About 1% unchanged anthraquinone was also found in the urine. Within 48 hours of oral administration of 5 mg radio-labelled anthraquinone/kg body weight to male rats, 42% had been excreted as unchanged anthraquinone, mainly in the faeces, and ca. 24% as free or conjugated 2-hydroxyanthraquinone, mainly in the urine (Bayer, 1985).

The distribution and excretion of progressive doses of anthraquinone (chemical specification not provided) was studied in male Fischer rats following single oral or intravenous doses. (^{14}C)-Anthraquinone (40 to 50 µCi/kg) was administered by intravenous injection at a dose of 0.35 mg/kg in DMSO and by gavage at doses of 0.35, 3.5, 35 and 350 mg/kg body weight in corn oil (5 ml/kg body weight). The urine and faeces were collected separately and the total radioactivity excreted determined after 6, 12, 24, 48, 72 and 96 hours. The animals were then killed and the total radioactivity in the various organs and tissues was determined. Independent of the route of administration and the dose, about $^1/_3$ of the total radioactivity was excreted in the urine and about $^2/_3$ in the faeces (see Table 3).

Table 3. Excretion and distribution of total radioactivity in Fischer rats 96 hours after a single intravenous or oral dose of ^{14}C-labelled anthraquinone (Winter et al., 1991, 1992; Winter, 1992)

Dose (mg/kg bw)	Data as % total radioactivity administered			Recovery rate in %
	Urine	Faeces	Body	
intravenous				
0.35	29.1	54.4	5.0	90
oral				
0.35	28.6	60.2	3.6	92
3.5	39.8	52.4	2.5	95
35.0	41.1	58.1	2.1	101
350.0	26.0	62.9	1.5	91

bw body weight

Only traces of radioactivity were found in the tissues (1.5 to 5.0%). After the animals were killed, the radioactivity was measured in the blood, brain, heart, kidney, liver, lungs, spleen, musculature, testes, skin, small intestine, large intestine, stomach, adipose tissue and in the contents of the stomach, small intestine and large intestine. The highest levels of radioactivity were found in the liver, kidneys and blood, 96 hours after intravenous or oral administration. The radioactivity in the gut was higher after intravenous injection than oral administration. This may be attributable to the enterohepatic circulation of anthraquinone and/or its metabolites. In order to determine the clearance of anthraquinone and its metabolites from the blood and the time dependence of the anthraquinone concentration in the tissues, animals were killed 1, 4, 24 and 96 hours after a single intravenous injection of 0.35 mg ^{14}C-anthraquinone/kg body weight and 4 and 96 hours after an oral dose of ^{14}C-anthraquinone (0.35 mg/kg body weight in corn oil). The amount of radioactivity was then measured in the major organs and tissues (blood, liver, kidney, adipose tissue, musculature, skin, brain, heart, lungs, spleen, testes, stomach, small and large intestines) and in the contents of the stomach, small intestine and large intestine. Within 1 to 4 hours, anthraquinone was distributed through all of the tissues. The highest concentrations were measured in the adipose tissue and the contents of the small and large intestines after 1 and 4 hours. Special analytical investigations of the adipose tissue revealed that only anthraquinone and not its metabolites were present there. Almost all of the radioactivity was eliminated from the body within 24 hours. There was no accumulation in any of the tissues. There was no difference between intravenous or oral administration with respect to the clearance of anthraquinone. The high excretion of radioactivity in the faeces of the rat following intravenous injection of radiolabelled anthraquinone supports the considerable significance of biliary excretion. This was confirmed in a study in rats fitted with biliary fistulae, as in these animals, 35% of the administered radioactivity was excreted in the bile within 6 hours of an intravenous infusion of 0.35 mg ^{14}C-anthraquinone/kg body weight. Radiochromatographic analysis of the bile samples revealed that cumulatively, less than 3% of the administered radioactive anthraquinone was eliminated as the unchanged product. The separation and identification of the anthraquinone metabolites excreted in the urine of rats treated with ^{14}C-anthraquinone showed numerous metabolites in the urine in all dose groups. Only 2-hydroxyanthraquinone and another metabolite, almost certainly 1-hydroxyanthraquinone, were identified (Steup et al., 1990; Winter et al., 1991, 1992; Winter, 1992).

The identification of the metabolites caused great difficulties. With HPLC, only ca. 60% of the radioactivity loaded on the column was recovered. This fact, and the knowledge that some of the metabolites of anthraquinone described in the literature rearrange in the aqueous phase, or are reduced or oxidised (e.g. anthrone to anthraquinone) has largely prevented attempts to identify the radioactive metabolite peaks with the aid of specific hydrolytic enzymes. New studies aimed at further structural elucidation are therefore being undertaken (Winter, 1992).

7.2   Acute and subacute toxicity

Acute toxicity

The results of all acute toxicity tests carried out in various species of laboratory animal and in sheep, following oral, intraperitoneal and dermal administration and inhalation exposure are shown in Tables 5, 6, 7 and 8. Details of the purity of the product tested were given in 8 study reports for a total of 14 of 27 acute toxicity tests. According to this information, the main products used were technical grade products with an anthraquinone content ranging from 97 to 98%. In these studies, anthraquinone was of very low toxicity, with only sheep showing a higher susceptibility. This is shown in the summary of acute toxicity studies with this substance (see Table 4).

Table 4. Summary of the results of acute toxicity tests on anthraquinone

Species	Route	LD_{50} g/kg bw after		
		1	7	14 days
Mouse	oral	> 5.0	> 5.0	> 5.0
Rat	oral	> 5.0	> 5.0	> 5.0
Rat	oral	> 10.0	> 10.0	>10.0
Sheep	oral	> 0.625	0.30	between 0.15 and 0.30 (lethal dose)
Mouse	intraperitoneal	> 6.81	> 6.81	> 6.81
Rat	intraperitoneal	> 5.0	> 5.0	> 5.0
Rat	dermal	> 5.0	> 5.0	> 5.0
Rabbit	dermal	> 3.0	> 3.0	> 3.0
Rat	inhalation*	>1327*	>1327*	>1327*

bw body weight
* LC_{50} maximum dust concentration that could be produced
     nominal: 1800, measured 1327 mg/m³, duration of exposure 4 hours

These results are discussed below, arranged according to the route of application.

Table 5. Acute toxicity of anthraquinone following oral administration by gavage

Strain	Sex	Specification	Vehicle	LD_{50} in g/kg bw after days			Findings			Reference
				1	7	14	Clinical	Post–mortem	Laboratory	
Studies in rats										
–	Male/female	97%, 0.5 – 0.7% fluorenone, 0.5 – 0.7% xanthone, <1.0% water	Water and carboxymethyl cellulose	–	>6.81	>6.81	Reduced body weight gain	No specific effects	–	BASF, 1974
–	Male	–	Water and Cremophor EL	>5.0	>5.0	>5.0	Slight reduction in general well–being (5 days), slight dyspnoea (2–5 days)	–	–	Bayer, 1975a
Wistar II	Male/female	98%	Water and Cremophor EL	>5.0	>5.0	>5.0	Slight reduction in general well–being (6 days), slight dyspnoea (2 days)	–	–	Bayer, 1975b
Albino Charles–River	Female	Crude, no further details	Aqueous suspension	>10.0	>10.0	>10.0	Reduced activity for 1 day	No specific effects	–	Industrial Bio–Test Laboratories, 1975d
–	Male	98%	Distilled water and Cremophor EL	> 5.0	> 5.0	> 5.0	Slight reduction in general well–being (2 days),	–	–	Bayer, 1977

–	Male	98.8% irradiated	Distilled water	> 5.0	> 5.0	> 5.0	diarrhoea –	–	–	Bayer, 1978c
Wistar II SPF	Male	"Dispersion"	Water	> 5.0	> 5.0	> 5.0	No specific effects	–	–	Bayer, 1979c
Wistar II	Male/ female	Preparation containing 25% anthraquinone	Distilled water and Cremophor EL	Based on Morkit 25 > 5.0 > 5.0 > 5.0 Based on anthraquinone content > 1.25 > 1.25 > 1.25			Slight reduction in general well–being (3 days), slight dyspnoea (3 days)	No specific effects	–	Bayer, 1974
–	–	–	–	15.0 No data on observation period			–	–	–	Izmerov et al., 1982
–	–	–	–	> 20.0 (lethal dose) No data on observation period			–	–	–	Marhold,1986

Studies in mice

NMRI	Male/ female	98%	Water and Cremophor EL	> 5.0	> 5.0	> 5.0	Slight reduction in general well–being (6 days), slight dyspnoea (2 days)	–	–	Bayer, 1975b

Studies in sheep (administered in the morning, before feeding, after fasting)

Black face	Male/ female	99.3%	Water and tylose	Lethal dose > 0.625	Between 0.30	0.15 and 0.30	Marked reduction in feed intake, weight loss	No specific effects	particular haemato–logical plasma urea and creatinine increased	Bayer, 1983a

–   no data
bw body weight

Oral administration (see Table 5)

The acute toxicity of anthraquinone following oral administration to mice and rats was consistently >5 g/kg body weight, independent of the vehicle used. In 3 studies, doses even higher than 5 g/kg body weight were administered and were survived: 10 g/kg body weight (Industrial Bio-Test Laboratories, 1975d), 15 g/kg body weight (Izmerov et al., 1982) and 20 g/kg body weight (Marhold, 1986). The publications of Izmerov et al. (1982) and Marhold (1986) contain no further information, and both of these LD_{50} values can thus only be evaluated to a limited extent.

In the acute toxicity studies, the animals showed mild, non-specific, reversible symptoms in the first days after administration only. Post-mortems, which were only carried out in 3 rat studies, revealed no abnormalities of the internal organs (BASF, 1974; Industrial Bio-Test Laboratories, 1975d; Bayer, 1974).

Sheep (2 animals/dose) were more sensitive than small rodents to anthraquinone, as late deaths occurred in this series of studies, and the lethal dose for this species was between 0.15 and 0.30 g/kg body weight after 14 days. A dose of 0.075 g/kg body weight was tolerated by both sheep without clinically, haematologically, biochemically or morphologically detectable effects. In animals treated with higher doses of anthraquinone, a dose-dependent reduction in feed intake occurred extending to complete refusal of feed and weight loss. In the 0.30 g/kg group, one of the 2 sheep died after 6 days and the second was killed on day 8 in a moribund condition. The animals treated with 0.625 g/kg died 3 and 14 days after anthraquinone administration, respectively. In the sheep treated with 0.15 g anthraquinone/kg body weight and above, pronounced increases in the plasma urea and creatinine concentrations were found. Kidney damage was therefore discussed by the investigators as a possible cause of the deaths of the animals. However, no effects on the organs were found at post-mortem, even in the fatally-poisoned sheep (Bayer, 1983a).

Intraperitoneal injection (see Table 6)

Anthraquinone was also of very low toxicity in rats and mice after intraperitoneal injection. The maximum doses injected (6.81 g/kg

Table 6. Acute toxicity of anthraquinone on intraperitoneal injection

Species Strain	Sex	Specification	Vehicle	LD_{50} in g/kg bw after days			Findings		Reference
				1	7	14	Clinical	Post–mortem	
Mouse	Male/ female	97.0%, 0.5-0.7% fluorenone, 0.5-0.7% xanthone, <1.0% water	Water and carboxymethyl-cellulose	> 6.81	> 6.81	> 6.81	Slight dyspnoea, mild apathy, fibrillary convulsions (transient)	Residual substance and adhesions in abdominal cavity	BASF, 1974
Rat	-	-	-	3.5 ± 0.6 No data on observation period			-	-	Volodchenko et al., 1971
Rat Wistar II	Male/ female	98%	Water and Cremophor EL	> 5.0	> 5.0	> 5.0	Slight reduction in general well-being (7 days), slight dyspnoea (5 days)	-	Bayer, 1975b
Rat	-	-	-	3.5 ± 0.6			-	-	Izmerov et al., 1982
Rat	-	-	-	20.0 (lethal dose) No data on observation period			-	-	Marhold, 1986

bw   body weight

body weight in mice, 5.0 g/kg body weight in rats) did not lead to deaths during the 14-day observation period, and the animals showed mild, non-specific clinical symptoms only during the first days after injection. In the mice, residues of the substance in the abdominal cavity and adhesions of the intestines were found at autopsy (at the end of the 14-day observation period). These types of effects are non-specific and are very often observed following intraperitoneal injection of substances that are not completely dissolved. In the publications by Volodchenko et al. (1971) and Izmerov et al. (1982) lower acute toxicity values are given (3.5 g/kg body weight). Information other than these figures was not provided for these rat studies.

Dermal application (see Table 7)

The application of 0.5 g (for 24 hours; Bayer, 1975b), 1.0 g (for 7 days; Bayer, 1972) or 5.0 g (for 24 hours; Bayer, 1982) anthraquinone/kg body weight to the shaved dorsal skin of rats under occlusive cover did not lead to deaths during the 14-day observation period and caused only mild, transitory impairment of the general well-being of the animals. Post-mortems on the rats treated with 5.0 g anthraquinone/kg body weight, which were carried out at the end of the 14-day observation period, revealed no pathological changes at the application site or in the internal organs. A 25% preparation of anthraquinone was also not toxic to rats under these study conditions (Bayer, 1974). Rabbits tolerated the application of 3.0 g anthraquinone/kg body weight to the scarified skin, with no pathological effects detectable at autopsy (for 24 hours; Industrial Bio-Test Laboratories, 1975a).

Inhalation (see Table 8)

In an inhalation hazard test, rats tolerated whole-body exposure to vapour saturated with anthraquinone at 20 °C for 8 hours without symptoms (BASF, 1974). In a further experiment, rats were exposed for 4 hours to the maximum anthraquinone concentration that could be produced with a dynamic dust generator, 1800 mg/m^3 air (head-nose exposure). Analytically, the concentration measured at

Table 7. Acute toxicity of anthraquinone after dermal application (dorsal skin)

Strain	Sex	Specification	Vehicle	24-hour occlusive cover	Shaving on previous day	LD_{50} in g/kg bw after days			Findings		Reference
						1	7	14	Clinical	Post-mortem	
Studies in rats											
Wistar II	Male	Technical	Water and Cremophor EL	+	+	> 1.0	> 1.0	> 1.0[1]	No specific effects	-	Bayer, 1972
Wistar II	Male/ female	98%	Water and Cremophor EL	+	+	> 0.5	> 0.5	> 0.5[1]	Slight reduction in general well-being	-	Bayer, 1975b
Wistar Bor-Wistar (SPF-Cpb)	Male/ female	98.2%	Physiological saline	+	+	> 5.0	> 5.0	> 5.0	Slight reduction in general well-being	No specific effects	Bayer, 1982
Wistar II	Male/ female	Preparation containing 25% anthraquinone	Distilled water and Cremophor EL	+	+	Based on Morkit 25 > 1.0 / Based on anthraquinone content > 0.25	> 1.0 / > 0.25	> 1.0 / > 0.25	No specific effects	-	Bayer, 1974
Studies in rabbits											
New Zealand White	Female	Crude, no further details	Water	+	Scarifi-cation	> 3.0	> 3.0	> 3.0	No specific effects	No specific effects	Industrial Bio-Test Laboratories, 1975a

1 higher doses not tested   −   no data   bw   body weight

Table 8. Acute inhalation toxicity of anthraquinone in rats

strain	Sex	Specification	Inhalation method	Exposure			LD_{50} in mg/m^3 after days			Findings		Reference
				Head/ nose	Whole body	Duration (hours)	1	7	14	Clinical	Post– mortem	
–	Male/ female	97%, 0.5 – 0.7% fluorenone, 0.5 – 0.7% xanthone, < 1% water	Vapour saturated with anthraquinone at 20 °C	Ø	+	8	No analysis, no mortality			No specific effects	No specific effects	BASF, 1974
Wistar Bor–Wis (SPF–Cpb)	Male/ female	98.2%	Dynamic dust generator	+	Ø	4	> 1327 Analytically determined	>1327	>1327	Non– specific symptoms (1 day)	No specific effects	Bayer, 1982
BLU: (LE)	Male/ female	Crude, no further	Dynamic dust generator	Ø	+	4	> 244	> 244	> 244	No specific effects	No specific effects	Industrial Bio–Test Laboratories, 1975c
–	–	–	–	–	–	6	> 12 No data on measurement or observation period	> 12	> 12	–	–	Izmerov et al., 1982
Wistar	Male/ female 25% anthraquinone	Preparation containing	Dynamic dust generator	+	Ø	1	Based on Morkit 25 > 700 Based on anthraquinone content > 175	> 700 > 175	> 700 > 175	No specific effects	–	Bayer, 1974
				+	Ø	4	Based on Morkit 25 > 750 Based on anthraquinone content >182.5	> 750 >182.5	> 75 >182.5	No specific effects	–	

– no data

the nose level of the animals was 1327 mg/m^3. Apart from non-specific behavioural disturbances on the day of exposure, the rats appeared normal. There were no treatment-related pathological effects on the organs in the animals killed and dissected at the end of the 14-day observation period (Bayer, 1982). Anthraquinone was also not toxic when inhaled as a 25% preparation for 1 or 4 hours under these study conditions (Bayer, 1974). Rats also tolerated a 4-hour whole-body exposure to an anthraquinone concentration of 244 mg/m^3 (which was produced with a dynamic dust generator) without signs of toxicity (Industrial Bio-Test Laboratories, 1975c). A 6-hour LC_{50} value of >12 mg/m^3 has been published in a handbook (Izmerov et al., 1982).

Subacute toxicity

Repeated oral administration

"Inhibition of the absorption and excretion function of the liver", which was studied with the Bengal red method, was described after daily oral administration (7 times in total) of 5 mg anthraquinone/kg body weight. In rats in which "hepatitis" had been produced by treatment with carbon tetrachloride, anthraquinone had no effect on the altered function of the liver (Pidemskii and Masenko, 1970). No data were provided on the purity of the anthraquinone used, the strain of animal or the type and frequency of the clinical, laboratory, pathological or anatomical studies carried out. There was also no information in the report on whether a control group was included in the study or whether there was a subsequent treatment-free observation period.

In a 30-day subacute study published by Volodchenko et al. (1971), anthraquinone-treated animals gained less weight than controls. Haematological tests revealed a slight reduction in haemoglobin and in the erythrocyte count, while clinical chemistry analyses revealed a tendency to albumin reduction and a decrease in γ-globulin. Light microscopy revealed effects on the liver, which were not further described and were said to be "a sign of protein dystrophy and haemodynamic disturbance". "Hyperplasia of certain lymphoid follicles" in the spleen and "swelling of the epithelium of the convoluted tubules" in the kidneys were mentioned by the authors, who regarded the changes in clinical chemistry and the effects seen by light microscopy as evidence that anthraquinone had a "definite hepatotoxic effect" (Volodchenko et al., 1971). This statement concerning a direct toxic effect on the liver caused by anthraquinone is in contrast to all other studies carried out with anthraquinone, in which the product was repeatedly administered in progressive doses to various species under clearly defined study conditions. For these reasons and because of the vague, unclear data in the publication of Volodchenko et al. (1971), the results are only of limited use for the overall evaluation of the toxicological profile of anthraquinone. Other experimental studies in animals are also reported in this publication, including the 30-day subacute study. It is not evident from this report how many animals of which species were used for which study. It is only reported that 96 rats, 10 rabbits and 50 mice were used in total for all of the studies. Furthermore, details of the route of exposure, the dose levels or the specification of the anthraquinone used are not provided. Based on the data on body weights given in the report (test animals 180±2 g, controls 225± 2.1 g), it is likely that the 30-day study was carried out in rats.

In two separate series of experiments, SPF-Wistar rats were treated with an oral dose of anthraquinone (as an emulsion in a Cremophor/water mixture) once daily on 28 consecutive days. No data were given on the chemical specification of the anthraquinone used. In the first study, the doses used were 0, 10, 50 and 250 mg/kg body weight, and in the second study they were 0, 2 and 20 mg/kg body weight; 15 males and 15 females were used per dose. The control rats were treated with the Cremophor/water mixture. A daily dose of 2 mg anthraquinone/kg body weight was tolerated by the rats without clinical, haematological or biochemical effects, or changes detectable by macroscopic examination or light microscopy, other than reduced adrenal gland weights in males and increased relative lung weights. The changes in organ weights were judged to be unrelated to treatment, as no dose-dependency or histopathological correlates existed. General malaise occurred from the 3rd week in the 10 and 20 mg/kg dose groups and from the second and first weeks in the 50 and 250 mg/kg dose groups, respectively. No deaths occurred in any of the dose groups. Body weight gain was decreased compared with the concurrent controls in females and males from daily doses of 20 and 50 mg/kg, respectively. The erythrocyte count was reduced in comparison with the controls at daily doses of 20 mg/kg and above. The concentrations of alanine aminotransferase (glutamate-pyruvate transaminase) and aspartate aminotransferase (glutamate-oxalate transaminase) were increased in the plasma at the top dose level (250 mg/kg body weight/day). At autopsy, dark stained spleens were evident in both sexes from 10 mg/kg. Histopathological examination revealed this to be due to increased blood supply to this organ. In addition, relative liver, kidney (females only) and spleen weights were increased from 10 mg/kg body weight/day and relative thyroid, heart, kidney and testis weights were increased in the males from 50 mg/kg body weight/day. Liver cell enlargement of varying degrees of severity and extent was diagnosed by light microscopy in all animals in the two highest dose groups (50 and 250 mg/kg) and in some of the animals in the 10 mg/kg group. None of the other organs examined showed any treatment-related microscopic effects. The increased liver weights and liver cell enlargement without concomitant degenerative effects are indicative of adaptation of the liver by stimulation of the enzymes that degrade foreign compounds. In this study, a daily dose of 2 mg anthraquinone/kg body weight was tolerated without adverse effects (Bayer, 1976).

Repeated dermal application

The application of a 15% solution of anthraquinone to the skin of rabbits on 30 occasions did not lead to any local effects (Volodchenko et al., 1971). The publication contained no further details on the specification of the anthraquinone used, the individual dosage, preparation of the solution, the strain of rabbit, the total duration of the study or any observation of the animals. This study is therefore only of limited use in risk assessment.

7.3   Skin and mucous membrane effects

Primary skin irritation in the rabbit (see Table 9)

7 samples of anthraquinone of differing specifications were tested for primary skin irritancy on the dorsal, flank or ear skin of the rab-

Table 9. Primary skin irritation of anthraquinone in the rabbit

Strain	Sex	Specification	Type of application	Site	Dose/ rabbit	Semi– occlusion (24 hours)	Washed off	Skin irritation after days				Reference
								Intact skin		Scarified skin		
								1	7	1	7	
–	–	Chemically pure	On cotton wool, dry or moistened	Ear	–	+	–	Ø	Ø	Not tested		Bayer, 1964
–	–	Crude with 2.9% anthracene		Ear	–	+	–	Ø	Ø			
–	–	Crude with 0.87% anthracene		Ear	–	+	–	Ø	Ø			
–	–	97.0%, 0.5 – 0.7% fluorenone, 0.5 – 0.7% xanthone, < 1.0% water	–	Back	–	–	–	Ø	Ø	Not tested		BASF, 1974
New Zealand White	Male/ female	Dispersion	–	Ear	0.5 ml	+	+	Ø	Ø	Not tested		Bayer, 1979b
New Zealand White	Male/ female	98.2%	Paste in water	Flank	0.5 g	+	–	Ø	Ø	Ø	Ø	Bayer, 1982
New Zealand White	Male/ female	Preparation containing 25% anthraquinone	On cellulose patch	Ear	0.5 g	+	–	Ø	Ø	Not tested		Bayer, 1974

Ø   no irritant effects   –   no data   +   carried out bit. Pure anthraquinone was used in 2 of the studies, while the others involved technical-grade anthraquinone containing various high levels of contaminants, such as anthracene. A repellent preparation containing 25% anthraquinone was also tested (Bayer, 1974). The substance was applied either dry or moistened (with water or oil) under semi-occlusive conditions for 24 hours. No effects were seen at the application site with either the pure anthraquinone or any of the various technical-grade products. Thus pure and technical-grade anthraquinone were not primary irritants to the rabbit skin (BASF, 1974; Bayer, 1964, 1974, 1979b, 1982).

Primary irritation of the mucous membranes of the rabbit eye
(see Table 10)

These studies were carried out with pure anthraquinone or with technical-grade anthraquinone containing various high levels of contaminants. The substance was introduced into the conjunctival sac of the rabbit as a powder, a dispersion or a solution. In some of the studies the eye was rinsed out 1 hour after introduction of the substance. In the majority of the studies, the effects on the eyes and the mucous membranes and on the conjunctiva were evaluated 1 hour, 1 day and 7 days after introduction of the substance, as well as, in one case (Bayer, 1982), 14 days after instillation of anthraquinone. Immediate sensory and inflammatory reactions seen on repeated daily introduction into the conjunctival sac rapidly disappeared (Estable, 1943). The author attributed these transitory, mild, local reactions to mechanical irritation by the powder, which is insoluble in the conjunctival fluid, rather than to a direct effect of anthraquinone. Corresponding observations were also made in other studies in which the product was introduced as a powder (BASF, 1974; Industrial Bio-Test Laboratories, 1975b; Bayer, 1982). No effects of any sort were seen in the studies in which the eyes were washed out after an hour (Bayer, 1964). Based on these studies, pure and technical-grade anthraquinone should be classified as not being primary irritants to the mucous membranes. A preparation that contained 25% anthraquinone was also not a primary irritant to the mucous membranes when introduced into the conjunctival sac of the rabbit (Bayer, 1974). No irritant effects were

Table 10. Primary irritant effect of anthraquinone on the mucous membranes of the rabbit eye

Strain of animal	Sex	Specification	Application Form	Dose	Rinsed out after 1 hour	Effects on the eye after				Reference
						1 hour	1	7 days	14	
–	–	–	powder	several times daily	no	(+)	Ø	–		Estable, 1943
–	–	chemically pure	powder	–	yes	Ø	Ø	Ø		Bayer, 1964
–	–	crude with 2.9% anthracene	powder	–	yes	Ø	Ø	Ø		Bayer, 1964
–	–	crude with 0.87% anthracene	powder	–	yes	Ø	Ø	Ø		Bayer, 1964
–	–	–	15% solution	–	–	Ø no times given				Volod-chenko et al., 1971
–	–	97%, 0.5 - 0.7% fluorenone, 0.5 - 0.7% xanthone, < 1.0% water	powder	50 mg	no	(+)	(+)	Ø		BASF, 1974
–	–	97%, 0.5 - 0.7% fluorenone, 0.5 - 0.7% xanthone, <1.0% water	talc	50 mg	no	(+)	(+)	Ø		BASF, 1974
Albino	–	crude, no further details	powder	100 mg	no	(+)	Ø	Ø		Industrial Bio-Test Laboratories, 1975b
–	–	dispersion	dispersion	100 µl	no	–	Ø	Ø		Bayer, 1979b
New Zealand White	♂ ♀	98.2%	powder	100 mg	no	(+)	(+)	Ø	Ø	Bayer, 1982
New Zealand White	♂ ♀	preparation with 25% anthraquinone	powder	50 mg	no	(+)	Ø	Ø		Bayer, 1974

Ø   non-irritant
(+) mild irritant
–   no data

described in the publication by Volodchenko et al. (1971) following the instillation of a 15% solution of anthraquinone into the conjunctival sac of the rabbit. No other data were provided and this study can therefore only be used to a limited extent in the evaluation of the primary irritant effect of anthraquinone on the mucous membranes (Estable, 1943; Bayer, 1964, 1974, 1979b, 1982; BASF, 1974; Industrial Bio-Test Laboratories, 1975b; Volodchenko et al., 1971).

Phototoxicity

In an initial series of experiments, the backs of nude mice were painted with a saturated solution of anthraquinone twice daily and then irradiated for 72 hours with an Osram-ultravitalux lamp (wavelength unspecified) from a distance of 50 cm. Skin erythema in the test mice was compared with that in nude mice in two control groups (one group only treated dermally with anthraquinone and one group irradiated only), immediately after irradiation and on the following days. In a second series of experiments, mice were given a daily dose of 100 mg anthraquinone/kg body weight by intraperitoneal injection and then exposed to UV light for 48 hours. The skin reaction in the anthraquinone-treated nude mice did not differ from that in the concomitant controls in either series of experiments. Thus, under the study conditions given, anthraquinone was not phototoxic to nude mice following dermal or intraperitoneal administration (Gloxhuber, 1970).

Following the application of a 15% solution of anthraquinone to the skin of rabbits with subsequent solar irradiation, hyperaemia of the treated skin was observed 30 to 60 minutes after irradiation. This was evaluated by the authors as a "weak photodynamic effect" (Volodchenko et al., 1971). Further data were not provided. This study is therefore only of limited value in the evaluation of the phototoxicity of anthraquinone.

Based on the study of Gloxhuber (1970) in nude mice, which was carried out under well-defined conditions, it is unlikely that anthraquinone has phototoxic properties. However, this cannot be completely excluded because of information in the literature relating to occupational medicine (see Sect. 8).

7.4   Sensitisation

In order to study skin sensitising potential in accordance with the recommendations of the US FDA (Draize, 1959), 14 female Pirbright guinea pigs (300 to 400 g) were given an induction treatment consisting of a total of 10 intracutaneous injections of a 0.5% solution of technical-grade (98% pure) anthraquinone 3 times weekly in the shaved skin of the flank. The volume injected on the first occasion was 0.05 ml and that on the following 9 occasions was 0.1 ml. As a challenge treatment the guinea pigs were given a further injection of 0.05 ml of a freshly prepared 0.1% anthraquinone solution 2 weeks after the 10th injection. The treated area was examined for redness and swelling 24 hours after this injection and the extent of the local effects was compared with the effects observed after the preceding 10 injections. Almost no local effects occurred after the challenge injection and technical-grade anthraquinone did not therefore induce skin sensitisation under these study conditions (Bayer, 1972).

7.5   Subchronic and chronic toxicity

Papillomas were seen in the gastric mucosa of rats following the daily oral administration of 0.2 mg anthraquinone for 90 days (no further details; Ito, 1941). These findings were not confirmed in later subchronic and chronic oral studies.

Groups of 20 male and 20 female Wistar rats, which weighed ca. 60 to 64 g at the beginning of the study, received technical-grade anthraquinone (98% pure) in the diet (powder) for 3 months at concentrations of 0 (controls), 15, 150 and 1500 ppm. Every week the rats were weighed and their feed consumption was determined. The average substance intake in mg/kg body weight calculated from the average feed consumption was somewhat lower in the males than in the females (see Table 11). Dose-dependent reductions in feed consumption were seen in the anthraquinone-treated rats compared with the controls, although the difference between the 15 ppm group and the control group was very slight. Haematological, biochemical and urine analyses were undertaken after 6 weeks and before the end of the study (after 3 months) in 5 males and 5

Table 11. Feed consumption and substance intake in a 3-month feeding study in rats

Dose ppm	Average feed consumption		Average substance intake	
	kg/rat[1]	g/rat/day	g/kg bw[1,2]	mg/kg bw/day
Male rats				
0	1.97	21.69	0	0
15	1.89	20.79	0.12	1.36
150	1.78	19.54	1.14	12.58
1500	1.69	18.52	11.49	126.25
Female rats				
0	1.66	18.20	0	0
15	1.60	17.58	0.16	1.79
150	1.49	16.35	1.53	16.79
1500	1.47	16.12	15.94	175.21

1) Average intake over the study period
2) Based on rat weight in the middle of the study

females from each treated and control group. No deaths occurred. After 3 months, all organs and tissues from the animals were evaluated macroscopically, organ weights were determined and the organs from 5 males and 5 females/group were examined by light microscopy. The rats in the 15 and 150 ppm groups gained less body weight than the controls during the first 3 weeks of the study. Thereafter, the body weight gain in the animals in these groups corresponded to that of the controls. These slight reductions in body weight gain, which in the two lower dose groups were only observed at the beginning of the study, were attributed by the authors to the repellent effect of the anthraquinone. Although body weight gain in the high dose rats was steady, it was markedly reduced compared with that in the controls. Apart from increased reticulocyte counts in isolated animals in the 150 and 1500 ppm groups, no haematological effects were evident compared with the controls. There were no changes in the clinical chemistry or urine analyses in the rats in the low and intermediate dose groups. In the rats in the 1500 ppm group, the serum cholesterol level was increased compared with the controls at the end of the study, indicating an impairment of cholesterol metabolism. The absolute liver weights of the rats in the 150 and 1500 ppm groups and the absolute thyroid gland weights in the males in the 1500 ppm group were increased dose-dependently and significantly compared with the controls. All other differences in organ weights were slight and not dose-related. The increased liver weights in the 150 and 1500 ppm groups and the enlarged centrilobular hepatocytes in the animals in the top dose group indicate an adaptation of the liver by stimulation of the enzymes that metabolise foreign compounds. No other treatment-related effects on the organs were established histopathologically. The temporary reduction in body weight gain accompanying reduced feed consumption in the low dose group (15 ppm), equivalent to 1.4 (male) and 1.8 (female) mg/kg body weight/day was thought to be of no toxicological relevance by the authors. A no effect level could not be determined in this study (Bayer, 1979a).

Rats were exposed to an anthraquinone-containing dust in chambers for 5 to 6 hours daily over 4 months in a "dynamic inhalation system". The anthraquinone content in the air was monitored by chemical analysis. The sensitivity of the method of analysis was 5 μg in 3 ml of the volume to be analyzed. Two concentrations were studied: 5.2 ± 0.8 mg/m^3 and 12.2 ± 0.7 mg/m^3. According to the authors, 12.2 ± 0.7 mg anthraquinone/m^3 was equivalent to the concentration measured most frequently under industrial conditions. As the rats were kept in chambers, the study probably involved whole-body exposure. Additional oral uptake of anthraquinone-containing dust cannot therefore be excluded. The lower concentration of 5.2 ± 0.8 mg dust/m^3 air was tolerated by the rats without clinically, haematologically or morphologically detectable effects or changes in clinical chemistry. A concentration of 5.2 mg anthraquinone/m^3 can therefore be considered to be a no effect level under these study conditions. In the rats exposed to the higher concentration of anthraquinone dust (12.2 ± 0.7 mg/m^3), transient changes occurred in comparison with the controls: reduced body weight gain and slight reductions in haemoglobin and in the erythrocyte count, linked with relative reticulocytopenia. In addition, lower vitamin C concentrations were determined in the blood of the anthraquinone-exposed animals compared with the controls during the first 3 months of the study. These effects returned to normal again during the 4th month of anthraquinone exposure. Light microscopic examination of the rats killed at the end of the 4-month inhalation period revealed non-specific effects on the lungs (disseminated emphysematous and atelectatic areas, disseminated exudation of fluid into the alveolar lumen and hyperaemia). These effects were no longer detectable in test animals killed after a 4-week observation period, and were thus reversible. The authors concluded that, considering the low toxicity of anthraquinone and observations in industry, a workplace guideline level of 10 mg/m^3 could be recommended (Volodchenko et al., 1971). The following data were not presented in the publication: specification of the anthraquinone used, strain of rat, number of animals in the test and control groups and number of rats/group during the observation period, description of the inhalation methods, frequency of laboratory studies and the type and number of organs and tissues per rat subjected to histopathological examination.

90-day oral studies carried out in F344 rats and B6C3F1 mice as part of the National Toxicology Program are currently being evaluated (NTP, 1994a).

7.6   Genotoxicity

7.6.1 In vitro

Studies in prokaryotes (see Table 12)

Anthraquinone has been tested for mutagenic activity by various investigators in the Salmonella/microsome test according to Ames et al. (1973a, b; McCann et al., 1975) and in some modifications of this test. The specifications of the anthraquinone used are given in 7 of a total of 16 publications and reports. Analytically pure anthraquinone was used by 5 of the groups and technical-grade anthraquinone by 2. The *Salmonella typhimurium* strains TA 1535 and TA 100, which are used to detect base-pair substitution mutations, were each used in 10 cases. In addition, a series of tests was carried out in the *Salmonella typhimurium* strain TA 1535 with the plasmid

Table 12. Mutagenic effect of anthraquinone in prokaryotes – Salmonella/microsome test (Ames test)

Specification	Salmonella typhimurium strains	Metabolic activation system (liver S9–mix)		Concentration µg/plate (solvent)	Mutagenic effect		Reference
		Species/strain	Inducer		without S9-mix	with S9-mix	
No data	TA 1535 TA 100 TA 1537 TA 98 TA 1538	Rat	Aroclor 1254, 1x100 mg/rat in corn oil	100 – 200 (DMSO)	Not mutagenic Not mutagenic Not mutagenic Not mutagenic Not mutagenic	Not mutagenic Not mutagenic Not mutagenic Not mutagenic Not mutagenic	Brown and Brown, 1976
No data	TA 1535 TA 100 TA 1537 TA 98 TA 1538	Not studied	Not studied	No data	Not mutagenic Not mutagenic Not mutagenic Not mutagenic Not mutagenic	Not studied Not studied Not studied Not studied Not studied	Brown et al., 1977
No data	TA 1535 TA 100 TA 1537 TA 98	Rat, male Sprague–Dawley	Aroclor 1254, 1x500 mg/kg i.p. in corn oil	4 – 2500 (DMSO)	Not mutagenic Not mutagenic Not mutagenic Not mutagenic	Not mutagenic Not mutagenic Not mutagenic Not mutagenic	Anderson and Styles, 1978
No data	TA 1535 TA 100 TA 1537 TA 98	Rate, male Sprague–Dawley	Aroclor 1254, 1x500 mg/kg i.p. in ground nut oil	3.7 – 300 (ethanol)	Not mutagenic Not mutagenic Not mutagenic Not mutagenic	Not mutagenic Not mutagenic Not mutagenic Not mutagenic	Bayer, 1978a
Industrial product, sieved	TA 1535 TA 100 TA 1537 TA 98	Rat, male Sprague–Dawley	Aroclor 1254, 1x500 mg/kg i.p. in ground nut oil	3.7 – 300 (ethanol)	Not mutagenic Not mutagenic Not mutagenic Not mutagenic	Not mutagenic Not mutagenic Not mutagenic Not mutagenic	Bayer, 1978b
Pure, analytically tested	TA 1535 TA 1537 TA 98 TA 1538	Not studied	Not studied	520 (DMSO)	Not mutagenic Not mutagenic Not mutagenic Not mutagenic	Not studied Not studied Not studied Not studied	Gibson et al., 1978

Pure, analytically tested, irradiated with ^{60}Co gamma (2.5 x10^7 rad)	TA 1535 TA 1537 TA 98 TA 1538	Not studied	Not studied	104 – 520 (DMSO)	Not mutagenic Not mutagenic Not mutagenic Not mutagenic	Not studied Not studied Not studied Not studied	Gibson et al., 1978
No data	TM 677 (8–Azaguanine–resistance)	Rat, male Sprague–Dawley Rat, male Sprague–Dawley	Aroclor 1254 Phenobarbital	20800 (=100 µM) 20800 (=100 µM)	Not studied Not studied	Not mutagenic Not mutagenic	Kaden et al., 1979
No data	TA 1535 TA 100 TA 1537 TA 98 TA 1538	Rat	Aroclor 1254	0.1 – 1000 (DMSO)	Not mutagenic Not mutagenic Not mutagenic Not mutagenic Not mutagenic	Not mutagenic Not mutagenic Not mutagenic Not mutagenic Not mutagenic	Salamone et al., 1979
No data	TA 1535 TA 100 TA 1537 TA 98 TA 1538	Rat	Aroclor 1254	0.2 – 20 (DMSO)	Not mutagenic Not mutagenic Mutagenic (from 10 µg) Mutagenic (from 10 µg) Mutagenic (from 2 µg)	Not mutagenic Not mutagenic Not mutagenic Not mutagenic Not mutagenic	Liberman et al., 1982
Pure, analytically tested	TA 100 TA 98 TA 2637	Rat or hamster	Phenobarbital and 5,6–benzoflavone	1 – 100 (DMSO)	Not studied Not studied Not studied	Not mutagenic Not mutagenic Not mutagenic	Tikkanen et al., 1983[1]
Pure, analytically tested	TA 100 TA 97 TA 98	Rat, male	PCB	5 – 200 (DMSO)	Not mutagenic Not mutagenic Not mutagenic	Not mutagenic Not mutagenic Not mutagenic	Sakai et al., 1985

(continued)

Table 12. (Continued)

Specification	Salmonella typhimurium strains	Metabolic activation system (liver S9–mix)		Concentration µg/plate (solvent)	Mutagenic effect		Reference
		Species/strain	Inducer		without S9-mix	with S9-mix	
97%, no data on contaminants	TA 100[1]	Rat, male Sprague–Dawley	Aroclor 1254	33 – 2500	Mutagenic (from 1000 µg)	Mutagenic (at 2500 µg)	Zeiger et al., 1988
		Golden hamster, male	Aroclor 1254	33 – 2500		Mutagenic (from 1000 µg)	
	TA 98[1]	Rat, male Sprague–Dawley	Aroclor 1254	33 – 2500	Mutagenic (from 33 µg)	Mutagenic (from 333 µg)	
		Golden hamster, male	Aroclor 1254	33 – 2500 (DMSO)		Mutagenic (from 333 µg)	
No data	TA 1535 pSK 1002 so–called umu test	Rat, male	Phenobarbital and 5,6–benzoflavone	680.3	Not mutagenic	Not mutagenic	Ono et al., 1991
Pure, analytically tested	TA 1537 TA 102	Rat, male Sprague–Dawley	Aroclor 1254	0.1 – 300	Not mutagenic Not mutagenic	Not mutagenic Not mutagenic	Krivobok et al., 1992
Pure (>99%)	TA 1535	Rat	Aroclor 1254	50 – 5000	Not mutagenic	Not mutagenic	Microbiological Associates, 1987
		Hamster	Aroclor 1254	50 – 5000	Not mutagenic	Not mutagenic	
	TA 100	Rat	Aroclor 1254	50 – 5000	Not mutagenic	Not mutagenic	
		Hamster	Aroclor 1254	50 – 5000	Not mutagenic	Not mutagenic	
	TA 1537	Rat	Aroclor 1254	50 – 5000	Not mutagenic	Not mutagenic	
		Hamster	Aroclor 1254	50 – 5000	Not mutagenic	Not mutagenic	
	TA 98	Rat	Aroclor 1254	50 – 5000	Not mutagenic	Not mutagenic	
		Hamster	Aroclor 1254	50 – 5000 (DMSO)	Not mutagenic	Not mutagenic	

1) Preincubation test

pSK 1002, which contains a fused gene (umuC⁻-lacZ). The umu-operon in this fused gene is induced by DNA-damaging substances. The extent of induction after DNA damage has taken place is measured by assay of β-galactosidase activity, which is expressed by the fused gene (Ono et al., 1991). In the frameshift marker strains, TA 1537, TA 98 and TA 1538, 11, 13 and 5 studies, respectively, were carried out with anthraquinone. One study in each case was carried out in the *Salmonella typhimurium* strains TA 97, TA 102, TA 2637 and the 8-azaguanine resistant strain TM 677 (Kaden et al., 1979). One group (Gibson et al., 1978) carried out tests in the *Salmonella typhimurium* strains TA 1535, TA 1537, TA 98 and TA 1538 with analytically pure anthraquinone and anthraquinone that had been γ-irradiated with cobalt 60 (2.5x10⁷ rad). Under these study conditions, neither the pure anthraquinone nor the irradiated anthraquinone was genotoxic. In 10 of the 16 studies, the tests were carried out with and without metabolic activation (by S9-mix). Two groups tested anthraquinone only for direct mutagenic effects (Brown et al., 1977; Gibson et al., 1978) and in two publications (Kaden et al., 1979; Tikkanen et al., 1983) tests for mutagenicity were carried out only following metabolic activation with S9-mix. The majority of investigators produced the S9-mix from the livers of Aroclor 1254-pretreated rats. Occasionally phenobarbital or 5,6-benzoflavone were used for induction and golden hamster liver was used for the production of S9-mix.

Anthraquinone caused no mutations in 13 of the 15 series of studies in the various *Salmonella typhimurium* strains. Only the group of Liberman et al. (1982) described a direct, although weak, mutagenic effect in strains TA 1537, TA 98 and TA 1538. Frameshift mutations did not occur in these strains following metabolic activation. This group did not obtain a positive effect in strains TA 1535 and TA 100, which are used to detect base-pair substitution mutations, either with or without the addition of S9-mix. The report contained no data on specifications. It is therefore not known whether pure or technical-grade anthraquinone was used for this test. Zeiger and co-workers (1988), who carried out tests in strains TA 100 and TA 98 using technical-grade anthraquinone (97% pure), obtained positive results in both Salmonella strains with and without metabolic activation. Data on the contaminants contained in the product were not reported. It is therefore possible that the mutagenic effect described by Zeiger and co-workers were not caused by anthraquinone, but by contaminants contained in the technical-grade product. The authors themselves indicated that it was not possible to elucidate whether the mutagenic effects detected with substances containing contaminants were caused by the test substances themselves or by the contaminants contained in the samples. Similar experimental experiences were described by Brown and Brown (1976) during the course of their studies of numerous amino-anthraquinone derivatives in the Salmonella microsome test.

Anthraquinone has been tested for mutagenicity in the Ames test not only as part of the National Toxicology Program (NTP; Zeiger et al., 1988), but also in the programme of the National Cancer Institute (NCI; Microbiological Associates, 1987). In preliminary NCI studies with and without metabolic activation in strain TA 100 with S9-mix from hamster liver, no cytotoxic or mutagenic effects were seen at concentrations of up to 2500 µg/plate. This result is contrary to the NTP studies (Zeiger et al., 1988), in which anthraquinone (97% pure) was mutagenic in the Salmonella strains TA 100 and TA 98. There were two differences in the Ames test protocols of the NTP and NCI. Under the NTP the Salmonella/microsome tests were carried out with the preincubation method with 30% hamster liver S9-mix, while the NCI standard protocol involved the plate incorporation test with 10% S9-mix. The preliminary studies carried out by the NCI with anthraquinone used the standard NCI method. Therefore, before the beginning of the main NCI study, further preliminary studies were undertaken in the Salmonella strain TA 100 with and without metabolic activation (with 30% S9-mix from hamster liver and the preincubation method). The anthraquinone concentrations used ranged from 3.33 to 2500 µg/plate. In contrast to the positive results with anthraquinone published by Zeiger et al. (1988), the NCI studies with anthraquinone using the preincubation method produced negative results. A further difference between the NTP and NCI studies was the purity of the anthraquinone used (NTP 97%, NCI >99%). As the additional NCI studies indicated that the difference in protocols could not be responsible for the differing test results between the preliminary NTP and NCI studies, the authors assumed that the different degrees of purity of the anthraquinone used were probably responsible for the differing results. The main NCI study was carried out with anthraquinone at concentrations of between 50 and 5000 μg/plate in accordance with the standard protocol of the NCI (plate incorporation test with 10% S9-mix). This pure anthraquinone did not cause any mutations in any of the tests either with or without metabolic activation with S9-mix from rat or hamster liver in the Salmonella strains TA 1535, TA 100, TA 1537 or TA 98 (Microbiological Associates, 1987).

Anthraquinone was also negative in the umu test in strain TA 1535 with the plasmid pSK 1002, a very sensitive assay used to detect the genotoxicity of by-products of chlorination and ozonisation, e.g. in water (Ono et al., 1991).

In summary, it has been ascertained that pure anthraquinone was not mutagenic in studies in various *Salmonella typhimurium* strains with or without the addition of S9-mix as a metabolic activation system. It can be assumed that in both of the studies with positive results, contaminants in the anthraquinone samples used were the cause of the mutagenicity detected.

According to Yamaguchi (1982), the mutagenic activity of two compounds which are mutagenic only after metabolic activation, namely Trp-P-2 (3-amino-1-methyl-5H-pyrido[2, 3-b]indole) and AAF (acetylaminofluorene), is considerably reduced if these

Table 13. Antimutagenic effect of anthraquinone on Trp-P-2 and AAF following metabolic activation by S9-mix from PCB-induced rat liver (Yamaguchi, 1982)

Mutagen	Indicator organism *Salmonella typhimurium*	Concentration μg/plate	Mutagenic effect number of revertants/plate[1]	
			S9-mix without anthraquinone	S9-mix with anthraquinone (12.5 μg)
Trp-P-2	TA 100	0.3	352	99
	TA 98	0.1	5526	93
AAF	TA 98	10.0	3616	165

Trp-P-2 = 3-Amino-1-methyl-5H-pyrido [2,3-b] indole
AAF   = Acetylaminofluorene
[1]   The average spontaneous revertant counts (40 in TA 98 and 120 in TA 100) were subtracted by the authors

mutagens are incubated with anthraquinone in the presence of S9-mix. In this experiment anthraquinone was added at a concentration of 12.5 μg/plate (see Table 13).

The reduction in mutagenicity of Trp-P-2 and AAF in the presence of anthraquinone was attributed to inhibition of the metabolic activation reaction of S9-mix. Other quinones tested in the course of this study also reduced the mutagenicity of Trp-P-2 and AAF by this mechanism (Yamaguchi, 1982).

7.6.2 In vivo

The cell nuclei from liver and kidney tissue of male mice (Swiss CD-1) were isolated and lysed on a filter membrane, 4 hours after intraperitoneal injection of the mice with 250 mg anthraquinone/kg (equivalent to 1.20 mmol/kg) body weight. The DNA was then eluted in an alkaline medium and the ratio of filter-bound to eluted DNA was determined. With this method, low-molecular-weight fragments elute more quickly than intact strands of high molecular weight. The cell nuclei from liver and kidney tissue from anthraquinone-treated mice showed an increased amount of low-molecular-weight fragments compared with those from the liver and kidney tissue of control animals. This was indicative of DNA strand break formation. The effect of anthraquinone under these study conditions was less than that of other carcinogenic substances studied for comparison, e.g. benzidine, 4-aminotoluene and N-nitroso-N-methylurea. According to the authors, DNA strand breaks can also be caused by non-specific toxic effects (Cesarone et al., 1982).

The effects of anthraquinone on the bone marrow of B6C3F1 mice were studied in a micronucleus test, in which 5 mice per dose (10 at the highest dose) were given intraperitoneal injections of anthraquinone in corn oil on 3 consecutive days. The animals were given no feed for 24 hours before the first injection. In each treatment group, 2000 polychromatic erythrocytes (PCE) were evaluated. 9,10-Dimethyl-1,2-benzanthracene (DMBA) was used as the positive control.

Table 14. Study of anthraquinone in the micronucleus test in mice following intraperitoneal injection on 3 consecutive days

Daily dose (mg/kg body weight)	Micronuclei-containing PCE/ 1000 PCE
2000	1.0±0.3
1000	1.3±0.4
500	1.5±0.3
0	2.2±0.5
12.5 DMBA	8.2±1.4

Anthraquinone was not mutagenic under the study conditions specified (NTP, 1994b).

Zeller and Häuser (1974) investigated the effect of a mercury-containing disinfectant and two combined agents with added lindane or anthraquinone on the chromosomes of mitotic cells in the root tip of germinating cereal seeds (barley, rye, wheat and oats). Anthraquinone was added to the seeds at 0.2 g/100 g, in accordance with the manufacturer's recommendations on the maximum amount of the product to be used in the disinfection of cereal seeds (200 g/100 kg cereal seed). Under these study conditions, anthraquinone caused no polyploidy in the root tip cells of the 4 types of cereal, and therefore was not mutagenic to the genome in barley, rye, wheat or oats (Zeller and Häuser, 1974).

7.7   Carcinogenicity

Short-term tests

Tests for liver regeneration

Anthraquinone was included in a study in which the growth-stimulating effects of aromatic hydrocarbons and their derivatives and of heterocyclic compounds were tested in partially (two-thirds) hepatectomised rats (Holtzmann or Charles-River). No data were given on the specification of the anthraquinone used in the study.

In each case treatment was administered after the partial hepatectomy. Anthraquinone was tested at 0.6% in the diet. Under these study conditions, anthraquinone significantly promoted liver regeneration. This effect should not be judged to be a possible indication of anthraquinone having carcinogenic activity, as these studies revealed no direct correlation between carcinogenic activity and stimulation of regrowth of the liver. In the same series of experiments, regrowth of the liver was not stimulated by the administration of known carcinogens such as 3,4-benzpyrene, 9,10-dimethyl-1,2-benzanthracene, 1,2,5,6-dibenzanthracene or benzene. Furthermore, the study revealed no clear correlation between the stimulation of liver enzyme inducing properties and liver growth in the treated, partially hepatectomised rats (Gershbein, 1975).

Anthraquinone was tested in 6 short-term tests which were being assessed for their usefulness in the prediction of the carcinogenic potential of organic chemicals (Purchase et al., 1978). The 6 tests were the Salmonella/microsome test (Anderson and Styles, 1978) and the cell transformation test (Styles, 1978), which gave the best overall agreement with the results of animal carcinogenicity studies (94% and 93%) and 4 methods (degranulation of the endoplasmic reticulum, sebaceous gland suppression, tetrazolium reduction and subcutaneous implantation) which did not give good correlations (Purchase et al., 1978). The results of the Salmonella/microsome test are described in Sect. 7.6, while those of the remaining tests are described below. Anthraquinone was negative in all of the studies, with the exception of the sebaceous gland suppression test.

Cell transformation test in mammalian cells in vitro

The cell transformation test was carried out in accordance with the method of Styles (1977). No details were given concerning the chemical specification of the anthraquinone tested. Two cell lines were used in the study: Chang (from human liver) and BHK-21C13 (from kidney cells of newborn golden hamsters). Anthraquinone was dissolved in DMSO at concentrations of 0.008, 0.04, 0.2, 1.0, 5.0 and 25 µg/ml. Quantities of these stock solutions were added to 1 ml of the cell suspension, so that the following concentrations were tested: 0.08, 0.4, 2.0, 10.0, 50.0 and 250.0 µg/ml. The spontaneous transformation rates in the surviving cells were 8 foci/10^6 cells in the Chang cell line and 55 foci/10^6 cells in the BHK-21C13 cell line. The LC_{50} values of the compounds were determined with these increasing concentrations, and thereafter the tests were carried out at the LC_{50} concentration with and without the addition of S9-mix from rat liver. The S9-mix was produced in accordance with the method described by Ames et al. (1975). For anthraquinone, the frequency of cell transformation per 10^6 surviving cells without and with the addition of S9-mix, respectively, was 11 and 13 foci/10^6 cells in the Chang cell line and 56 and 39 foci/10^6 cells in the BHK-21C13 cell line. Thus anthraquinone did not cause cell transformation under these study conditions. Most of the carcinogens studied under these conditions tested positive (Styles, 1978).

Degranulation test

Degranulation of the rough endoplasmic reticulum (RER) has been described as an early lesion, following the treatment of animals with carcinogens. The consequence is an increase in the smooth endoplasmic reticulum (SER). The radiolabelled cell membranes required for the degranulation test were taken from the livers of rats given intraperitoneal injections of 6-^{14}C-labelled orotic acid monohydrate. The difference between the specific activity of the SER and the RER represents a measure of the degranulation. No increased degranulation relative to the controls was detected after incubation of the membrane with anthraquinone (12 µg/ml) dissolved in DMSO for 2 hours (Lefevre, 1978).

Sebaceous gland suppression test

The dorso-lumbar region of 55-day-old SPF mice of the Swiss-derived Alderley Park strain were clipped and then treated twice daily with 0.1 ml of the test solution (in DMSO) on 3 consecutive days. The ratio of sebaceous glands to hair follicles was then quantitatively determined by histological examination. Anthraquinone (chemical specification not given) caused a <50% reduction in sebaceous glands and thus was weakly positive in this test according to the classification given by the author. A reduction in sebaceous glands compared with the controls of 50 to 74% was judged to be clearly positive, while a decrease of >75% was classified as strongly positive. Contradictory results were obtained with the sebaceous gland suppression test. It is therefore not suitable as a short-term test for carcinogenicity (Longstaff, 1978a).

Tetrazolium reduction test

Anthraquinone was tested during the course of a study of 118 compounds in a modified tetrazolium reduction test. The substances were applied to the skin of mice either in solution or in suspension. After 2 days the animals were killed and the treated area of skin was incubated with colourless tetrazolium red solution. The formation of coloured formazan was then determined spectrophotometrically. Anthraquinone was negative in this test, as a similar amount of formazan was present in the area of skin treated with anthraquinone as in the skin of the control mice (Westwood, 1978).

Implantation test

Alderley Park Swiss mice (10 males and 10 females) were given subcutaneous implants in the dorso-lumbar region of discs containing 0.02 mmol (= 4.164 mg) anthraquinone. The physical properties of the implanted discs (which were 13 mm in diameter and had a pore size of 0.22 μm) were known not to induce tumours. Mice surviving 3 months after implantation were killed and examined for local and internal tumours. No tumours developed in the area of implantation in the 20 mice treated with anthraquinone in this manner. The histological appearance of the tissue surrounding the implanted disc corresponded to that in the control mice, which had discs without added substances implanted (Longstaff, 1978b).

Carcinogenicity tests

Oral administration

In a series of experiments with various substances, 18 male and 18 female mice of each of 2 hybrid strains (B6AKF1 and B6C3F1; 72

mice in total) were given anthraquinone. All of the mice were treated with daily doses of 464 mg anthraquinone/kg body weight by gavage from day 7 after birth until the age of 4 weeks. The dose was kept constant for 4 weeks and not adapted to increases in body weight. The vehicle used was a 0.5% aqueous gelatine suspension. After the mice were weaned from the dams at the age of 4 weeks, they were given a diet containing anthraquinone at a concentration of 1206 ppm for a further 17 months. All mice that died during the study and those surviving until the end were dissected and all major organs (except for the brain) and any macroscopically visible lesions were examined by light microscopy. Blood smears were prepared from all mice before sacrifice, but were only examined in animals that had spleen tumours, enlarged livers or lymphadenopathy. No increase in the tumour incidence was found in the anthraquinone-treated mice in comparison with the various concomitant control groups. Anthraquinone was thus not carcinogenic under the study conditions given (Innes et al., 1969).

Dermal application

The dorsal skin of mice (strain unspecified, approximately 1 month old) was painted daily or every other day with 0.1 or 0.25% solutions in benzene of quinone, α-naphthoquinone, quinone and α-naphthoquinone, anthraquinone, phenanthrenequinone, thymoquinone, chloroquinone and β-naphthoquinone. The control mice were treated with p-phenylenediamine in benzene or with benzene alone in the same manner (no data were provided on the number of animals/dose group at the beginning of the study, the frequency of skin painting or the specification of the test chemicals). Many mice died during the study. Although 1159 days was given as the longest survival time for one of the mice, the skin tumours that occurred were only evaluated in relation to the mice still alive in each group after 200 days. Data on the skin tumour incidence after a study duration of more than 200 days were not given in the report.

The results obtained with quinone, α-naphthoquinone and phenanthrenequinone were judged to be an indication of carcinogenic potential. In contrast, the occurrence of only 1 papilloma in

Table 15. Skin tumours in mice following repeated dermal application

Substance	Number of mice		
	alive after 200 days	with papillomas	with carcinomas
Quinone	87	9	3
α-Naphthoquinone	77	14	3
Quinone and α-naphthoquinone	46	7	3
Anthraquinone	38	1	0
Phenanthrenequinone	36	4	0
Thymoquinone	28	1	0
Chloroquinone	14	0	0
β-Naphthoquinone	25	0	0
p-Phenylenediamine	9	0	0
Benzene	46	1	0

each group after treatment with anthraquinone, thymoquinone or chloroquinone was considered not to be biologically relevant. Lung tumours were seen in various groups in the mice living longer than 200 days: 8/87 in the quinone group, 2/46 in the quinone and α-naphthoquinone group, 2/38 in the anthraquinone group, 2/14 in the chloroquinone group, 2/76 in the α-naphthoquinone group and 1/46 in the benzene group. These lung tumours, some of which had the morphology of adenocarcinomas, were not included in the evaluation by the authors (Takizawa, 1940).

Subcutaneous injection

The carcinogenic effect of anthraquinone (specification not provided) was tested in a subcutaneous injection study. The carcinogens dibenz(a,h)anthracene (administered subcutaneously or dermally) and 20-methylcholanthrene (administered dermally) served as positive controls. For each substance, 15 male and 15 female mice were used. Only one of the 30 mice treated subcutaneously with anthraquinone developed a tumour. In contrast, 10/30 mice treated subcutaneously with dibenzanthracene, 15/30 treated dermally with dibenzanthracene and 17/30 treated dermally with 20-methylcholanthrene developed tumours. Anthraquinone was therefore not carcinogenic under the study conditions given (Tada et al., 1966). The doses, duration of application, intervals between applications, total study duration and observation period are not evident from the English summary or the tables in English in this Japanese work.

The results of this study can only be judged as an indication that anthraquinone is not carcinogenic.

Suckling mice (18 male and 18 female, ca. 28 days old) of 2 hybrid strains (B6AKF1 and B6C3F1) were given a single subcutaneous injection (in the neck area) of 1000 mg anthraquinone/kg body weight as a 0.5% aqueous gelatine suspension. The melting point of the anthraquinone used was between 286 and 288 °C. The mice were observed for a total of 18 months. Blood smears were prepared one day before the end of the study, but were only examined in animals that had spleen tumours, enlarged livers or lymphadenopathy. The mice were dissected, the injection site and the internal organs examined for tumours and selected tissues and organs were subjected to histopathological examination. The controls were treated and examined in a corresponding manner. Anthraquinone was not carcinogenic under the study conditions specified (Bionetics, 1968).

Evaluation

The carcinogenicity studies available for anthraquinone are limited to 4 early reports; one study involving repeated dermal application, two involving subcutaneous injection and one subchronic oral study. No carcinogenicity tests which comply with current testing guidelines have been carried out with anthraquinone. Despite the limitations of the 4 early negative carcinogenicity studies, it can be assumed that pure anthraquinone probably has no carcinogenic potential as the pure substance is not genotoxic and the relevant short-term test (cell transformation test according to Styles, 1977, 1978) was negative.

7.8   Reproductive toxicity

In vitro teratogenicity studies in embryonic Drosophila cultures are based on the fact that teratogenic effects can be caused by abnormal cell death, defective cell interaction, reduced biosynthesis or impeded morphogenetic movement. The Drosophila cell cultures were kept for 24 hours at 26 °C and anthraquinone was tested at a concentration of 1000 μM. After exposure for 24 hours, the cell cultures were stained and the number of differentiated myotubes and ganglia were counted. In contrast to known tumour promoters that were studied at the same time, such as phenobarbital and phorbol esters, anthraquinone caused no reduction in the total number of myotubes or ganglia in comparison with control cultures. Anthraquinone did not therefore interfere with normal cell differentiation in embryonic Drosophila tissue. The study gave no indications of possible teratogenic effects (Bournias-Vardiabasis and Flores, 1986), although of course it can provide no insight into the teratogenic potential of anthraquinone in mammals.

7.9   Effects on the immune system

No information available.

7.10   Neurotoxicity

No information available.

7.11   Other effects

Biochemical studies

The binding of the carcinogen 3'-methyl-4-dimethylaminoazobenzene (3'-MeDAB) to rat liver protein is inhibited by certain metal chelating agents that have an o-hydroxybenzoic acid structure. 5 different anthraquinone derivatives were therefore tested and compared with anthraquinone. Male albino rats (weight 250 to 350 g) were given intraperitoneal injections of 82.5 mg 3'-MeDAB/kg body weight in groundnut oil, of groundnut oil alone or of 3'-MeDAB and the substance to be tested, one of which was anthraquinone (equimolar dose to 82.5 mg 3'-MeDAB/kg body weight). The animals were killed 24 hours later when binding to the liver protein was determined together with the glutathione (GSH) content of the liver. Anthraquinone inhibited the binding of the azo dye to the liver protein to the same degree as the other anthraquinone derivatives tested. Furthermore, anthraquinone caused a reduction in the GSH content in the liver. In contrast, the anthraquinone derivatives caused an increase in GSH (Neish and Key, 1966).

Anthraquinone slightly inhibited the activity of deoxyribonuclease from bovine pancreas in vitro at a concentration of 10^{-5} M (Hoffmann-Ostenhof and Frisch-Niggemeyer, 1952).

The activity of sodium and potassium activated ATPase activity from rabbit erythrocyte membranes was also inhibited by anthraquinone in vitro. Maximum inhibition of sodium-potassium ATPase was achieved with 5 mM anthraquinone. The enzyme inhibition caused by anthraquinone increased with increasing potassium concentration at a constant sodium concentration in the medium and with increasing sodium concentration at a constant potassium concentration in the medium. The inhibitory effect of anthraquinone on sodium/potassium ATPase activity was reduced by calcium ions. There was no relationship between the inhibitory effect of anthraquinone on the enzyme and the amino groups of lysine, the hydroxyl group of threonine and the imidazole group of histidine. Anthraquinone probably inhibits sodium/potassium ATPase activity through the sufhydryl or carboxyl groups of the enzyme (Koh, 1977).

Prostaglandin biosynthesis in methylcholanthrene-transformed 3T3 mouse fibroblasts in vitro was inhibited by 50% at an anthraquinone concentration of 2.4 μM. In studies with anthracene, phenanthrene and 7,8-benzoflavone, it was found that the inhibition of prostaglandin biosynthesis was reversible after removal of the substances (Levine and Hong, 1977).

In contrast to various quinone derivatives, anthraquinone (at a concentration of 10^{-5} M) barely stimulated the formation of superoxide anion radicals (O_2^-) by the three flavoprotein enzymes (NADPH-cytochrome P-450 reductase, NADH-cytochrome b5-reductase and NADH-ubiquinone oxidoreductase) in isolated rat hepatocytes (Powis et al., 1981).

Because of the selective inhibition of guanylate cyclase activity (GCA) by the cardiotoxic anthracycline antibiotic doxorubicin in rat and human hearts in vitro, anthraquinone and anthraquinone derivatives were tested for their inhibitory effect on this enzyme (which was obtained from the hearts of male Sprague-Dawley rats), at a concentration of 1 mM in vitro. Under these conditions, anthraquinone inhibited GCA activity only marginally compared with the control (20%; Lehotay et al., 1982).

When menadione (2-methyl-1,4-naphthoquinone) is used as a substrate, carbonyl reductase (a cytosolic, monomeric oxidoreductase with high specificity for carbonyl compounds) was shown to be the most important NADPH-dependent quinone reductase in the human liver. The purified enzyme catalyzed the reduction of a large number of natural and synthetic quinones. The activity measured with menadione as a substrate was designated as 100%, equivalent to 2.3 U/mg protein. On addition of anthraquinone (concentration tested 15 μmol), a relative enzyme activity of 22% was determined compared with menadione. The best substrates were found to be benzo- and naphthoquinones with short substituents (Wermuth et al., 1988).

Tests for antineoplastic effects

Powell (1944) studied the tumour-inhibiting effect of anthraquinone on Twort carcinomas in a study in which tumour-carrying mice were given intraperitoneal injections of 0.5 ml of a 0.1% anthraquinone solution twice daily for 13 days. In comparison with the untreated tumour-carrying controls, anthraquinone inhibited the growth of the Twort carcinoma by 46.9% (Powell, 1944).

The antineoplastic effect of anthraquinone on subcutaneous NF murine sarcomas was studied in vitro. Pieces of 10- to 14-day old sarcomas measuring ca. 1 mm in diameter were placed in petri dishes with solutions of anthraquinone (in physiological saline) at concentrations of 0.05, 0.01 and 0.005% at a pH of 5 to 6 and a temperature of 4 to 7 °C for 24 hours. The tumour pieces were then implanted subcutaneously in mice at 4 different sites and the growth of the sarcomas was monitored for 2 weeks. At each anthraquinone concentration 3 mice were used for tumour inoculation.

Anthraquinone did not inhibit the growth of the transplanted tumours under these study conditions (Sakai et al., 1955).

Single-cell suspensions were prepared from 24 malignant human ovarian tumours removed during surgery and the cytotoxicity of 10% dimethyl sulfoxide (DMSO) and each of 6 antineoplastic compounds (doxorubicin 0.04 µg/ml, anthraquinone 0.1 µg/ml, cis-platin 0.2 µg/ml, 5-fluorouracil 6 µg/ml, methomezate 100 µg/ml and vinblastin 6 µg/ml) alone was tested on them as well as the combination of DMSO with each of the cytostatic compounds. Exposure was at 37 °C for one hour. Anthraquinone alone was lethal to 15 and 33% of the tumour cells, respectively, in the two cell suspensions studied. The cytotoxicity increased to 53 and 51%, respectively, through the synergistic effect of DMSO. Even more pronounced synergistic effects were found with the other tested cytostatic compounds through the addition of DMSO. According to the authors it is probably therefore therapeutically advantageous to treat malignant ovarian tumours by intraperitoneal injection of combined cytostatics and DMSO (Pommier et al., 1988).

Pharmacological studies

Various K vitamins were tested for their anti-haemorrhagic activity in chicks fed vitamin K-free diets and compared with chemically-related compounds and other substances, such as anthraquinone. One chick was treated with 0.055 and a second with 0.136 mg anthraquinone/kg body weight/day (route and duration of application not specified). The reciprocal prothrombin activity was decreased from 10 and 11 to 3.5 and 1.1, respectively. Under the study conditions employed, the anti-haemorrhagic effect of anthraquinone was weaker than that of vitamin K. The specification of the anthraquinone tested was not provided (Dam et al., 1940).

Anthraquinone (specification not provided) had a sedative effect in mammals (species unspecified) following a single oral dose of 1 mg/kg body weight, but had no analgesic effect in mice, rats or guinea pigs after a single subcutaneous injection of 1 mg/kg body weight. A test for antipyretic activity in rabbits with hay infusion was also negative after the administration (probably oral, but not clearly stated) of 2 mg/kg body weight (Stern et al., 1957).

Anthraquinone was included in an investigation of 30 anthracene and diphenylmethane derivatives for their effect on water absorption from the colons of fasted rats. No significant difference in the water content of the gut lumen was found one hour after the injection of anthraquinone (10^{-3} mol) into the tied-off colon in vivo, compared with that after a control injection of Tyrode solution. Anthraquinone did not therefore affect water absorption from the colon of the rat (de Witte et al., 1991).

Anthraquinone was also tested in comparative studies of the cytotoxicity of polymer and mineral dusts to rat peritoneal and pulmonary macrophages in vitro, and of the fibrogenic activity of these materials in rats following intraperitoneal injection. Suspension cultures of rat peritoneal and pulmonary macrophages, which contained 10^6 cells/ml, were incubated for 2 hours with 0.5 mg dust/10^6 cells. The diameter of the anthraquinone particles ranged from 2 to 50 μm. To determine the cytotoxicity, vital staining with trypan blue was undertaken in 1 ml samples of the suspensions both before the addition of the dust or anthraquinone and 1 and 2 hours afterwards. Highly cytotoxic dusts such as chrysotile led to the deaths of more than 10% of the peritoneal and more than 60% of the pulmonary macrophages, while moderately cytotoxic compounds caused the deaths of 4 to 8% of peritoneal and 10 to 50% of pulmonary macrophages. Those compounds killing less than 2 and 5% of peritoneal and pulmonary macrophages, respectively, which included anthraquinone and coal, were classified as weakly cytotoxic. In the animal studies, groups of 6 male and 6 female Wistar rats (Alderley Park strain) were given intraperitoneal injections of the dust suspension (in physiological saline) at 50 mg/kg body weight and the tissue reactions were assessed 1 and 3 months later by gross examination and by light microscopy. The liver, spleen, pancreas and omentum were also examined. In contrast to asbestos (chrysotile) and other substances, which caused granulomas and, after 3 months, diffuse fibrosis in the abdominal cavity of the rats, anthraquinone, and also coal and pelican ink, caused no tissue reactions at all. Under the study conditions employed, anthraquinone was therefore not cytotoxic in vitro and not fibrogenic in vivo (Styles and Wilson, 1973).

The effect of anthraquinone on energy metabolism in human skin fibroblasts (Flow 7000 from forehead skin) was studied by microcalorimetric measurement of the evolution of heat from fibroblast cultures during the various phases of their growth in vitro compared with untreated cell cultures. At a concentration of 15 μM, anthraquinone did not reduce the production of heat and thus energy metabolism and viability of the fibroblasts, and was not therefore cytotoxic under these study conditions (Pätel and Antipolis, 1981).

8.   Experience in humans

Hyperpigmentation of the skin of the face and neck developed in a 37-year-old man who was sporadically occupationally exposed to anthraquinone, aniline, naphthalene, benzene, toluene, phthalic acid anhydride, benzanthrone, chlorobenzanthrone, tetraline, alizarin, o-benzoylbenzoic acid, various mineral oils, sulfur dioxide, sulfur trioxide, various alkalis and sodium bisulfite. To elucidate the cause, all of these substances, including anthraquinone, were applied to the abdominal skin of the patient in two skin tests. In the direct test, the compounds were dissolved in petrol and applied to an area of skin measuring half a square inch (equivalent to 1.27 cm^2). After the solvent had been evaporated, the remaining substances were covered with gauze for 18 hours. In the irradiation test, the substances were applied to the skin in the same manner, but the skin was irradiated with UV light at a sub-erythematous dose immediately before being covered. Anthraquinone did not cause any visible effects on the skin, such as inflammatory reactions or increased pigmentation, in either test and was thus not a primary irritant nor a phototoxin under these study conditions (Wieder, 1932).

During the production of anthraquinone in a contact process, contamination of the air in the working areas with anthraquinone dust occurred. Concentrations measured were up to 10 mg/m^3 at the contact apparatus, between 55 and 840 mg/m^3 at the balance and up to 1650 mg/m^3 during cleaning of the gas line. The workers involved in the anthraquinone production complained of headaches, general malaise and irritation of uncovered skin and eyes (Volodchenko et al., 1971).

A 40-year-old man who was employed in processing cellulose pulp in a paper factory suffered from recurrent subacute dermatitis of the exposed skin (face, neck and backs of the hands). The dermatitis recurred each time a new batch of anthraquinone was added to the paper pulp and always disappeared when work was stopped. It always recurred on return to the area of the factory in question and became more severe on exposure to sunlight. Simple patch tests carried out with 2 commercial samples of anthraquinone were negative. Photo-patch tests, in which these anthraquinone samples were irradiated with UVA, UVB or visible light, were positive only after irradiation of the anthraquinone with UVA. Histologically, eczematous changes were diagnosed in biopsy material from the areas of skin giving positive reactions. Photo-patch tests carried out in 5 control subjects with the same samples of anthraquinone were negative. Despite the negative results in the control tests and the histological diagnosis of eczema, the authors assumed, because of the monomorphic clinical picture with oedema and erythema, the absence of exudative lesions, the confinement of effects to the exposed sites and the weak reaction in the photo-patch test, that the effects seen in the worker were a phototoxic reaction due to anthraquinone. The 99% pure commercial anthraquinone used contained various contaminants: 0.3% phenanthrene, 0.05% anthracene, 0.3% anthrone, 0.1% nitrobenzene and 0.3% secondary volatile components, some of which could have had phototoxic effects. The contaminants could therefore have contributed to the development of the photodermatitis in this worker (Brandao and Valente, 1988).

Anthraquinone was identified as the main contaminant in the accumulated dust in a factory manufacturing catalytic anthraquinone in which women aged from 25 to 40 years were the main employees. In the female workers who had been employed at the factory for between 1 and 4 years, an increase in the latent period of direct and indirect reflexes to visual and auditory stimuli was observed as well as a reduction in stamina, particularly towards the end of the shift. The changes in these parameters were interpreted by the authors as a functional impairment of the central nervous system. Other effects observed in the workers during the shift included increases in the pulse frequency (by 3 to 10 beats/minute), the maximum arterial blood pressure (by 5 to 13 mmHg) and the skin temperature of the body (by 0.7 to 1.4 °C) and hands (by 0.8 to 2.0 °C).

These effects were most pronounced in condenser workers and those involved in packing the end product (Labunskii et al., 1976).

Bernier (1979) studied protein loss in the gastro-intestinal tract of volunteers by measurement of the plasma clearance of α-1-antitrypsin before and for 3 days after a single oral dose of anthraquinone (0.20 g) or other laxatives (magnesium sulfate (10 g), phenolphthalein (0.20 g), castor oil (10 g) and lactulose (33 to 66 g). Under physiological conditions, protein loss through the gastrointestinal tract in humans is 20 to 40 g per day measured by the direct method (60 to 100 ml from the plasma, 12 to 33 g in digestive secretions and by loss of 50 milliard cells from the gastrointestinal tract). Protein loss from the plasma was increased 4- to 10-fold by a single dose of the laxatives magnesium sulfate, anthraquinone, phenolphthalein and castor oil. Under these study conditions, lactulose did not affect protein loss from the plasma of the volunteers (Bernier, 1979).

9.   Threshold limit values

No information available.

References

Ames, B.N., Lee, F.D., Durston, W.E.
An improved bacterial test system for the detection and classification of mutagens and carcinogens
Proc. Natl. Acad. Sci. (USA), 70, 782–786 (1973a)

Ames, B.N., Durston, W.E., Yamasaki, E., Lee, F.D.
Carcinogens are mutagens: a simple test system combining liver homogenates for activation and bacteria for detection
Proc. Natl. Acad. Sci. (USA), 70, 2281–2285 (1973b)

Ames, B.N., McCann, J., Yamasaki, E.
Methods for detecting carcinogens and mutagens with Salmonella/ mammalian-microsome mutagenicity test
Mutat. Res., 31, 347 (1975)

Anderson, D., Styles, J.A.
The bacterial mutation test
Br. J. Cancer, 37, 924–930 (1978)

BASF AG, Medizinisch-Biologische Forschungslaboratorien,
Gewerbehygiene und Toxikologie
Gewerbetoxikologische Vorprüfung
Unpublished report no. XXIV33 (1974)

Bayer AG, Toxikologisches und Gewerbehygienisches Laboratorium
Anthrachinon – Hautfunktionsprüfungen
Unpublished report (1964)

Bayer AG, Institut für Toxikologie
Anthrachinon – Toxikologische Untersuchungen
Unpublished report no. 3802 (1972)

Bayer AG, Institut für Toxikologie
Morkit 25 Trockenbeize – Toxikologische Untersuchungen
Unpublished report no. 4909 (1974)

Bayer AG, Institut für Toxikologie
Anthrachinon Eg. 1/75 – Bestimmung der akuten Toxizität (LD_{50})
Unpublished report (1975a)

Bayer AG, Institut für Toxikologie
Anthrachinon – Untersuchungen zur akuten Toxizität
Unpublished report no. 5287 (1975b)

Bayer AG, Institut für Toxikologie
Anthrachinon – Untersuchung zur subakuten Toxizität an Ratten bei
oraler Applikation (4-Wochen-Versuch)
Unpublished report no. 5806 (1976)

Bayer AG, Institut für Toxikologie
Anthrachinon Eg. 2/72, 98 %ig – Bestimmung der akuten Toxizität (LD_{50})
Unpublished report (1977)

Bayer AG, Institut für Toxikologie
CH 002 – Salmonella/Microsome-Test zur Untersuchung auf
punktmutagene Wirkung
Unpublished report no. 7590 (1978a)

Bayer AG, Institut für Toxikologie
CH 001 – Salmonella/Microsome-Test zur Untersuchung auf
punktmutagene Wirkung
Unpublished report no. 7622 (1978b)

Bayer AG, Institut für Toxikologie
Anthrachinon (gestrahlt) Eg. 1/78; 98,8%ig – Bestimmung der
akuten Toxizität (LD_{50})
Unpublished report (1978c)

Bayer AG, Institut für Toxikologie
Anthrachinon – Subchronische toxikologische Untersuchungen an
Ratten (Fütterungsversuch über 3 Monate)
Unpublished report no. 8169 (1979a)

Bayer AG, Institut für Toxikologie
Untersuchung zur Haut- und Schleimhautverträglichkeit
Unpublished report (1979b)

Bayer AG, Institut für Toxikologie
Anthrachinon-Dispersion – Akute orale Toxizität
Unpublished report (1979c)

Bayer AG, Institut für Toxikologie
Anthrachinon (Techn.) – Untersuchungen zur akuten Toxizität
Unpublished report no. 11333 (1982)

Bayer AG, Institut für Toxikologie
Anthrachinon-technisch – Akute Toxizität am Schaf nach oraler
Verabreichung
Unpublished report no. 11761 (1983a)

Bayer AG, Sparte Pharma
[9, 10-^{14}C]Anthrachinon (Morkit®): Biokinetische Untersuchungen
an Ratten
Unpublished report no. 12013 (F) (1983b)

Bayer AG, Geschäftsbereich Pflanzenschutz
Biotransformation von [9, 10-^{14}C]Anthrachinon bei Ratten nach
oraler Applikation
Unpublished report no. 2393 (1985)

Bayer AG
AIDA-Grunddatensatz 9,10-Anthracenedion (1993a)

Bayer AG
Personal communication (1993b)

Bernier, J.J.
Les pertes protéiques digestives. Un effet secondaire des laxatifs
Bull. Acad. Natl. Med., 163 (8), 825–831 (1979)

Bionetics Research Laboratories, Inc., Bethesda, Maryland, USA
Evaluation of carcinogenic, teratogenic, and mutagenic activities of selected pesticides and industrial chemicals. Volume I. Carcinogenicity study
Report of contract nos. PH 43–64–57 and PH 43–67–735 (1968)
On behalf of the National Cancer Institute
NTIS/PB-223 159

Bournias-Vardiabasis, N., Flores, J.C.
Response of Drosophila embryonic cells to tumor promoters
Toxicol. Appl. Pharmacol., 85, 196–206 (1986)

Brandao, F.M., Valente, A.
Photodermatitis from anthraquinone
Contact Dermatitis, 18 (3), 171–172 (1988)

Brown, J.P., Brown, R.J.
Mutagenesis by 9,10-anthraquinone derivatives and related compounds in *Salmonella typhimurium*
Mutat. Res., 40, 203–224 (1976)

Brown, J.P., Dietrich, P.S., Brown, R.J.
Frameshift mutagenicity of certain naturally occurring phenolic compounds in the Salmonella/microsome test: activation of anthraquinone and flavonol glycosides by gut bacterial enzymes
Biochem. Soc. Trans., 5, 1489–1492 (1977)

Budavari, S., O'Neil, M.J., Smith, A., Heckelman, P.E. (eds.)
The Merck index
11th ed., p. 109
Merck & Co., Inc., Rahway, N.J., U.S.A. (1989)

Cesarone, C.F., Bolognesis, C., Santi, L.
Evaluation of damage to DNA after in vivo exposure to different classes of chemicals
Arch. Toxicol., Suppl. 5, 355–359 (1982)

Cofrancesco, A.J.
Anthrachinone
In: Kroschwitz, J.I., Howe-Grant, M. (eds.)
Kirk-Othmer encyclopedia of chemical technology, p. 801–814
John Wiley and Sons, Inc., New York, USA (1991)

Copplestone, J.F.
Pesticides
In: Parmeggiani, L. (technical ed.)
Encyclopaedia of occupational health and safety
Vol. 2, p. 1616–1633
International Labour Office (ILO), Geneva (1983)

Dam, H., Glavind, J., Karrer, P.
Die biologische Aktivität der natürlichen K-Vitamine und einiger verwandter Verbindungen
Chim. Acta, 23, 224–233 (1940)

Draize, J.H.
Dermal toxicity
In: Appraisal of the safety of chemicals in foods, drugs and cosmetics by the staff of the division of pharmacology Food and Drug Administration Department of Health, Education and Welfare, p. 46–59
Association of Food and Drug Officials of the United States, Business Office, Bureau of Food and Drugs, Texas State Department of Health, Austin 1, Texas (1959)

Estable, J.J.
The ocular effect of several irritant drugs applied directly to the conjunctiva
Ophthalmology, 31, 837–844 (1943)

Gershbein, L.L.
Liver regeneration as influenced by the structure of aromatic and heterocyclic compounds
Res. Commun. Chem. Pathol. Pharmacol., 11 (3), 445–466 (1975)

Gibson, T.L., Smart, V.B., Smith, L.L.
Non-enzymatic activation of polycyclic aromatic hydrocarbons as mutagens
Mutat. Res., 49, 153–161 (1978)

Gloxhuber, C.
Prüfung von Kosmetik-Grundstoffen auf fototoxische Wirkung
J. Soc. Cosmet. Chem., 21, 825–833 (1970)

Hoffmann-Ostenhof, O., Frisch-Niggemeyer, W.
Über die Beeinflussung der Aktivität der Desoxyribonucleinase aus Rinderpankreas durch Chinone und Phenole
Monatshefte für Chemie und verwandte Teile anderer Wissenschaften, 83, 1175–1179 (1952)

Industrial Bio-Test Laboratories, Inc.
Acute dermal toxicity study – albino rabbits
Unpublished report (1975a)
On behalf of the Ciba Geigy Corp., USA

Industrial Bio-Test Laboratories, Inc.
Eye irritation test – albino rabbits
Unpublished report (1975b)
On behalf of the Ciba Geigy Corp., USA

Industrial Bio-Test Laboratories, Inc.
Acute dust inhalation toxicity study with A. Q. rec. dry in rats
Unpublished report (1975c)
On behalf of the Ciba Geigy Corp., USA

Industrial Bio-Test Laboratories, Inc.
Acute oral toxicity study – female albino rats
Unpublished report (1975d)
On behalf of the Ciba Geigy Corp., USA

Innes, J.R.M., Ulland, B.M., Valerio, M.G., Petrucelli, L., Fishbein, L., Hart, E.R., Pallotta, A.J., Bates, R.R., Falk, H.L., Gart, J.J., Klein, M., Mitchell, I., Peters, J.
Bioassay of pesticides and industrial chemicals for tumorigenicity in mice: a preliminary note
J. Natl. Cancer Inst., 42, 1101–1114 (1969)

Ito (1941)
Cited in: NTP (1988)

Izmerov, N.F., Sanotsky, I.V., Sidorov, K.K.
Toxicometric parameters of industrial toxic chemicals under single
exposure, p.22 (English translation of the Russian)
Centre of International Projects, GKNT, Moscow (1982)

Kaden, D.A., Hites, R.A., Thilly, W.G.
Mutagenicity of soot and associated polycyclic aromatic hydrocar-
bons to *Salmonella typhimurium*
Cancer Res., 39, 4152–4159 (1979)

Koh, I.S.
Action of anthraquinone on sodium-potassium activated ATPase in
rabbit red cell membrane (Korean with English summary)
Taehan Saengri Hakhoe Chi, 11, 1–9 (1977)

Krivobok, S., Seigle-Murandi, F., Steiman, R., Marzin, D.R., Betina,
V. Mutagenicity of substituted anthraquinones in the Ames/Salmo-
nella microsome system
Mutat. Res., 179, 1–8 (1992)

Labunskii, V.V., Artemenko, G.E., Gudz, Z.A., Dobzhinskii, V.G.
Hygienische Charakteristik der Arbeitsbedingungen bei der Herstellung
von katalytischem Anthrachinon (German translation of the Russian)
Tr. Khark. Med. Inst., 124, 63–65 (1976)

Lefevre, P.A.
Appendix IV. The degranulation test
Br. J. Cancer, 37, 937–943 (1978)

Lehotay, D.C., Levey, B.A., Rogerson, B.J., Levey, G.S.
Inhibition of cardiac guanylate cyclase by doxorubicin and some of
its analogs: possible relationship to cardiotoxicity
Cancer Treat. Rep., 66, 311–316 (1982)

Levine, L., Hong, S.L.
Analogues of anthracene, phenanthrene, and benzoflavone inhibit
prostaglandin biosynthesis by cells in culture
Prostaglandins, 14, 1–9 (1977)

Liberman, D.F., Fink, R.C., Schaefer, F.L., Mulcahy, R.J., Stark, A.A.
Mutagenicity of anthraquinone and hydroxylated anthraquinones in the Ames/Salmonella microsome system
Appl. Environ. Microbiol., 43, 1354–1359 (1982)

Longo, R.
Attività e metabolismo nel topo di componenti la corteccia di Rhamnus frangula e di prodotti sintetici a struttura anthrachinonica
Boll. Chim. Farm., 119, 669–689 (1980)

Longstaff, E.
Appendix V. The sebaceous-gland test
Br. J. Cancer, 37, 944–948 (1978a)

Longstaff, E.
Appendix VII. The implant test
Br. J. Cancer, 37, 954–959 (1978b)

Marhold, J.
Prehled Prumyslové Toxikologie, Organické Latky, p. 299 and 300 (Review of industrial toxicology of organic compounds; translation of the Czech)
Avicenum, zdravotnické nakladatelstvi, Prague (1986)

McCann, J., Spingarn, N.E., Kobori, J., Ames, B.N.
Detection of carcinogens as mutagens: bacterial tester strains with R factor plasmids

Proc. Natl. Acad. Sci. (USA), 72, 979–983 (1975)
Microbiological Associates Inc.
Salmonella/mammalian-microsome plate incorporation mutagenicity assay (Ames test)
Unpublished report, study no. C73.501017 (1987)
On behalf of the National Cancer Institute
Also reported, in abstract form, in: CCRIS (Chemical Carcinogenesis Research Information Service), produced by NCI (National Cancer Institute), Bethesda, Maryland, USA

Munn, A., Smagghe, G.
Dyes and dyestuffs
In: Parmeggiani, L. (technical ed.)
Encyclopaedia of occupational health and safety
Vol. 2, p. 699–701
International Labour Office (ILO), Geneva (1983)

Neish, W.J.P., Key, L.
Inhibition of aminoazo dye binding to rat liver protein by anthra-
quinone derivatives
Biochem. Pharmacol., 15, 2127–2129 (1966)

NTP (National Toxicology Program), National Institute of Health,
Bethesda, Maryland, USA
Draft report: executive summary of data – anthraquinone, October
31, 1988

NTP (National Toxicology Program)
Written communication to BG Chemie of 29.04.1994a

NTP (National Toxicology Program)
Written communication to BG Chemie of 27.05.1994b

Ono, Y., Somiya, I., Kawamura, M.
The evaluation of genotoxicity using DNA repairing test for chemi-
cals produced in chlorination and ozonation processes
Wat. Sci. Tech. (Kyoto), 23, 329–338 (1991)

Pätel, M., Antipolis, S.
Calorimetric screening test for dermatologically active drugs on
human skin fibroblast-cultures
Thermochim. Acta, 49, 123–129 (1981)

Pidemskii, E.L., Masenko, V.P.
Effect of some antiinflammatory preparations on the absorption and
excretory function of a normal liver and one affected by experimen-
tal hepatitis
Tr. Perm. Gos. Med. Inst., 99, 325–328 (1970)
Abstract from: TOXALL (TOXLINE)
Produced by National Library of Medicine (NLM), Bethesda, Mary-
land, USA

Pommier, R.F., Woltering, E.A., Milo, G., Fletcher, W.S.
Synergistic cytotoxicity between dimethyl sulfoxide and antineo-
plastic agents against ovarian cancer in vitro
Am. J. Obstet. Gynecol., 159, 848–852 (1988)

Powell, A.K.
Growth-inhibiting action of some pure substances
Nature, 153 (3881), 345 (1944)

Powis, G., Svingen, B.A., Appel, P.
Quinone-stimulated superoxide formation by subcellular fractions,
isolated hepatocytes, and other cells
Mol. Pharmacol., 20, 387–394 (1981)

Purchase, I.F.H., Longstaff, E., Ashby, J., Styles, J.A., Anderson, D.,
Lefevre, P.A., Westwood, F.R.
An evaluation of 6 short-term tests for detecting organic chemical
carcinogens
Br. J. Cancer, 37, 873–903 (1978)

Salamone, M.F., Heddle, J.A., Katz, M.
The mutagenic activity of thirty polycyclic aromatic hydrocarbons
(PAH) and oxides in urban airborne particulates
Environ. Int., 2, 37–43 (1979)

Sakai, S., Minoda, K., Saito, G., Fukuoka, F.
On the anti-cancer action of quinone derivatives
Gann, 46, 59–66 (1955)

Sakai, M., Yoshida, D., Mizusaki, S.
Mutagenicity of polycyclic aromatic hydrocarbons and quinones
on *Salmonella typhimurium* TA97
Mutat. Res., 156, 61–67 (1985)

Sato, T., Fukuyama, T., Yamada, M., Suzuki, T.
Metabolism of anthraquinone. I. Isolation of 2-hydroxyanthraqui-
none from the urine of rats
J. Biochem., 43 (1), 21–24 (1956)

Sato, T., Suzuki, T., Yoshikawa, H.
Metabolism of anthraquinone. II. Sulfate conjugate of 2-hydroxyan-
thraquinone
J. Biochem., 46 (8), 1097–1099 (1959)

Sims, P.
Metabolism of polycyclic compounds. 25. The metabolism of
anthracene and some related compounds in rats
Biochem. J., 92, 621–631 (1964)

Stern, P., Dezelic, M., Kosak, R.
Über die analgetische und antipyretische Wirkung von k-Vitamin und
Dicumarol mit besonderer Berücksichtigung des 4-Hydroxycumarins
Naunyn Schmiedebergs Arch. Pharmacol., 232, 356–358 (1957)

Steup, M.B., Winter, S.M., Sipes, I.G.
Distribution and excretion of anthraquinone in the male F-344 rat
Toxicologist, 10 (1), 240, abstr. no. 957 (1990)

Styles, J.A., Wilson, J.
Comparison between in vitro toxicity of polymer and mineral dusts
and their fibrogenicity
Ann. Occup. Hyg., 16, 241–250 (1973)

Styles, J.A.
A method for detecting carcinogenic organic chemicals using mam-
malian cells in culture
Br. J. Cancer, 36, 558–563 (1977)

Styles, J.A.
Appendix III. Mammalian cell transformation in vitro
Br. J. Cancer, 37, 931–936 (1978)

Tada, K., Odashima, N., Ishidate, M.
On the screening experiment for the carcinogenesis of polycyclic
quinones
(Japanese with English summary)
Kyoritsu Yakka Daigaku Kenkyo Nempo, 5, 63–68 (1966)

Takizawa, N.
On the carcinogenic action of certain quinones (preliminary report)
Proc. Imperial Acad., 16, 309–312 (1940)

Tikkanen, L., Matsushima, T., Natori, S.
Mutagenicity of anthraquinones in the Salmonella preincubation test
Mutat. Res., 116, 297–304 (1983)

Vogel, A.
Anthraquinone
In: Ullmann's encyclopedia of industrial chemistry
5th ed., vol. A2, 347–354
Verlag Chemie, Weinheim (1985)

Volodchenko, V.A., Gudz, Z.A., Tomchenko, A.N.
Material zur Begründung der maximal zulässigen Anthrachinon-Grenzkonzentration in der Luft eines Arbeitsraumes (German translation of the Russian)
Gig. Tr. Prof. Zabol., 15 (2), 58–59 (1971)

Wermuth, B., Platts, K.L., Seidel, A., Oesch, F.
Carbonyl reductase provides the enzymatic basis of quinone detoxication in man
Biochem. Pharmacol., 35, 1277–1282 (1988)

Westwood, F.R.
Appendix VI. The tetrazolium-reduction test
Br. J. Cancer, 37, 949–953 (1978)

Wieder, L.M.
Occupational melanosis
Arch. Dermatol. Syphilol., 25, 624–643 (1932)

Winter, S.M., Steup, M.B., Sipes, I.G.
Distribution and excretion of anthraquinone in the male Fischer-344 rat
Toxicologist, 11, 90, abstr. no. 273 (1991)

Winter, S.M., Kattnig, M.J., Steup, M.B., Sipes, I.G.
Tissue distribution and excretion of anthraquinone in the male Fischer-344 rat
Toxicologist, 12, 163, abstr. no. 576 (1992)

Winter, S.M.
The metabolism and disposition of anthraquinone in the male Fischer 344 rat
Preliminary results (1992)

Witte, de, P., Dressen, M., Lemli, J.
The influence of some anthracene and diphenylmethane derivatives on water and electrolyte movement in rat colon
Pharm. Acta Helv., 66 (3), 70–73 (1991)

Yamaguchi, T.
Reduction of induced mutability with biologically active quinones through inhibition of metabolic activation
Agric. Biol. Chem., 46 (9), 2373–2375 (1982)

Zeiger, E., Anderson, B., Haworth, S., Lawlor, T., Mortelmans, K.
Salmonella mutagenicity tests: IV. Results from the testing of 300 chemicals
Environ. Mol. Mutagen., 11, Suppl. 12, 1–158 (1988)

Zeller, F.J., Häuser, H.
Polyploidisierung von Getreidearten durch Lindan-haltige Beizmittel
Experientia, 30 (4), 345–348 (1974)

o-Phthalodinitrile

This Toxicological Evaluation replaces the previously published version in volume 2

1. Summary and assessment

o-Phthalodinitrile is toxic on acute exposure by the oral, intraperitoneal and subcutaneous routes (LD$_{50}$ rat oral ca. 30 to 150 mg/kg body weight, mouse oral 65.1 to 100 mg/kg body weight, rat intraperitoneal 50 to 62 mg/kg body weight, mouse intraperitoneal ca. 5 to 50 mg/kg body weight, mouse subcutaneous ca. 7.94 to 55 mg/kg body weight). It is of low toxicity on dermal application to rabbits for 20 hours (LD$_{50}$ >5000 mg/kg body weight). On acute exposure, o-phthalodinitrile can cause disturbance of balance, augmented reflexes, convulsions and paralysis in all species of laboratory animal, independent of the route of exposure, except on acute dermal exposure in rabbits, when no systemic toxic effects are observed. Similar signs of systemic toxicity are observed on repeated dermal application. On repeated oral administration of doses that are not toxic on acute exposure, no signs of cumulative effects are seen. In a dose-finding study which included tests for neurotoxicity, the administration of 5, 15 or 45 mg o-phthalodinitrile/kg body weight/day in the diet (nominal doses) for 14 days led to reductions in body weight gain that were slight in males in the intermediate dose group and significant and accompanied by reduced feed intake in both sexes in the high dose group during the entire study period. No effects on histopathological, biochemical or haematological parameters were found.

o-Phthalodinitrile is not irritating to the skin or eyes of the rabbit, but is slightly irritating to the skin of the guinea pig.

Mutagenic effects have not been detected in any of the strains of *Salmonella typhimurium* tested in the Ames test, either with or without activation. o-Phthalodinitrile has also given no indications of mutagenic activity in the HPRT test in V79 cells or of genotoxic activity in a micronucleus test in mice.

In a long-term study, the increased incidence of leukaemia in mice on oral, subcutaneous and dermal exposure and in rats after oral and subcutaneous exposure to the product for 2 years can only be evaluated to a limited extent. Details of concomitant control groups and the cause of death of the numerous rats and mice that were not evaluated are not provided in the study report, and the documentation of individual findings is inadequate. Norpoth (1983) has suggested that the initial administration of toxic doses with cumulative effects could possibly have led to activation of latent leukaemia viruses in the mice, leading to the increased incidence of leukaemias.

No macroscopic pathological effects on the central or peripheral nervous system have been determined in post-mortems on rats after intraperitoneal administration of o-phthalodinitrile for 1 or up to 90 days. Histopathologically, degenerative effects on the nerve fibres in the brain, particularly in the hypothalamus and the underlying thalamic nucleus, were only detectable in the high dose groups (single doses of 50 or 70 mg/kg body weight or subchronic treatment with 10 mg/kg body weight/day) with Nauta-Gygax staining (silver impregnation). The value of these results is limited, as this staining method was only used in 1 to 3 rats/group, and no morphological effects were found with other staining methods, even at higher doses. In a 90-day neurotoxicity study in Sprague-Dawley rats, the 15 males/group were given 3, 8 or 25 mg o-phthalodinitrile/kg body weight/day and the 15 females/group 3, 10 or 30 mg/kg body weight/day in the diet. To test for possible cumulative or delayed effects, 5 rats/sex/group were subjected to interim post-mortems after 28 days. A thorough neurotoxicological study (Functional Observational Battery), including quantitative determination of motor activity, was carried out before the study began and in weeks 3, 7 and 12 of the study. With the aid of a perfusion technique, 5 animals/sex/group underwent a thorough, specific neuropathological study. In this subchronic study, o-phthalodinitrile caused reductions in body weight gain compared with the controls during the entire study period, effects which correlated with significantly reduced feed consumption. These effects were significant in both sexes in the high dose group and were also seen in females in the intermediate dose group where they did not achieve statistical significance. Qualitative and quantitative neurotoxicity tests revealed that o-phthalodinitrile caused a significant increase in activity rates in a dose- and time-dependent manner in females in the intermediate and high dose groups as early as the third week, and in males in the high dose group only, but only in week 12 of the study. No macromorphological or neurohistopathological correlates were found when specific staining methods were used in the different parts of the central and peripheral nervous system. Clouding of the lens was detected in eye examinations at the end of the study in both sexes in the high dose group and in some females in the intermediate dose group, an effect that was not evident after 4 weeks. Thus in this study, the main effects of subchronic oral exposure to o-phthalodinitrile were reduced body weight gain accompanying reduced feed intake, increased levels of activity and clouding of the lens of the eye in both sexes in the high dose group and to some extent in the females in the intermediate dose group. The no observed adverse effect level in this study was thus 3 mg/kg body weight/day.

In humans involved in the manufacture of o-phthalodinitrile, cases of acute intoxication have been described following absorption through the skin and inhalation of the dust; heavy perspiration and prolonged skin contact favour percutaneous absorption which, after exposure to relatively small amounts and after a latent period of hours, has led to severe, epileptiform convulsions lasting minutes, with disturbances of consciousness, bradycardia and, in some cases, retrograde amnesia. The convulsive attacks have either occurred without preliminary symptoms in the midst of activity or following symptoms such as dizziness, nausea, vomiting and headaches. No pathological effects were seen in a histological examination of the brain carried out immediately after a poisoning which proved fatal due to fracture of the skull in the course of a convulsive attack. There were also no clinical or biochemical deviations from the norm or pathological effects on EEGs in subjects who had experienced o-phthalodinitrile intoxication. In 21 workers who were employed in a plant producing o-phthalodinitrile for an average of $1^1/_2$ years during the period between 1957 and 1959, there were reductions in the haemoglobin content and the erythrocyte count in the blood with increasing length of employment, together with a reduction in the average body weight compared with normal values. This was not confirmed in another study of 81 workers who had been exposed for an average of 8.5 years, 11 of whom had suffered acute o-phthalodinitrile intoxication. Clinical examination, neurological studies, laboratory studies of clinical chemistry and haematological parameters and EEG studies detected no abnormal findings in these workers. Studies of the chromosomes of industrially-exposed workers revealed no findings that were significantly different from those in a control group. There were no increases in the numbers of deaths or in malignant diseases in an epidemiological mortality study in workers who were exposed to o-phthalodinitrile.

2.   Name of substance

2.1   Usual name   o-Phthalodinitrile

2.2   IUPAC-name   1,2-Benzodicarbonitrile

2.3   CAS-No.   91–15–6

2.4   EINECS-No.   202–044–8

3.   Synonyms, common and trade names

1,2-Benzenedicarbonitrile
o-Benzenedicarbonitrile
o-Benzenedinitrile
1,2-Benzodinitrile
1,2-Bis(cyano)benzene
o-Cyanobenzonitrile
1,2-Dicyanobenzene
o-Dicyanobenzene
o-PDN

Phthalic acid dinitrile
o-Phthalodinitril
Phthalodinitrile

4.   Structural and molecular formulae

4.1   Structural formula

4.2   Molecular formula $C_8H_4N_2$

5.   Physical and chemical properties

5.1	Molecular mass, g/mol	128.12	
5.2	Melting point, °C	141–142	(decomposition) (Thiess, 1968)
		139.9	(Matsushita, 1981)
5.3	Boiling point, °C	–	
5.4	Vapour pressure, hPa	0.04 (at 20 °C)	
		0.5 (at 100 °C)	(BASF, 1993)
5.5	Density, g/cm³	1.24 (at 20 °C) (Weast, 1977/78)	
5.6	Solubility in water	Slightly soluble ca. 0.51 g/l (at 20 °C) 0.56 g/l (at 25 °C) (BASF, 1993)	
5.7	Solubility in organic solvents	Low in ligroin (Weast, 1977/78) and polyethylene glycol (Yoshikawa and Kawai, 1966) Dissolves well in alcohol, ether, chloroform (Weast, 1977/78) Dissolves very well in benzoni- trile (Norpoth, 1983)	

27 g/100 g acetone (at 25 °C)
5 g/100 g benzene (at 25 °C)
12 g/100 g nitrobenzene (at 25 °C)
5 g/100 g pyridine (at 25 °C)
13 g/100 g quinoline (at 25 °C)
3 g/100 g toluene (at 25 °C)
2 g/100 g xylene (at 25 °C)
4.7 g/100 g 95% ethanol
    (at 35 °C)   (Matsushita, 1981)
3% in ethanol (at 25 °C)
                    (Oesch, 1978)

5.8   Solubility in fat

5 g/100 g neutral oil (at 30 °C)
0.5 g/100 g olive oil (at 35 °C)
                  (Matsushita, 1981)

5.9   pH value

–

5.10 Conversion factor

1 ml/m^3 (ppm) ≙ 5.23 mg/m^3
1 mg/m^3 ≙ 0.19 ml/m^3 (ppm)
(at 1013 hPa and 25 °C)

6.   Uses

Intermediate in the manufacture of phthalocyanine pigment dyes and pesticides; stabiliser for aircraft fuels and in the rubber industry (Kleinsorge et al., 1979).

Starting substance in the manufacture of optical brighteners and sensitisers for photography (Towae et al., 1992).

7.   Experimental results

7.1   Toxicokinetics and metabolism

No information available.

7.2   Acute and subacute toxicity

Acute toxicity

The results of acute toxicity studies are shown in Table 1.

Table 1. Acute toxicity of o-phthalodinitrile in various species by different routes

Species	Route (preparation)	LD_{50} (mg/kg body weight) or dose at which effects seen	Reference
Mouse	oral (in polyethylene glycol)	65.1	ISK, 1991
Mouse	oral (in polyethylene glycol)	65.2	Yoshikawa and Kawai, 1966
Mouse	oral (in gum arabic)	100	Cheav and Foussard-Blanpin, 1990
Mouse	intraperitoneal (in Tween 20)	ca. 5-10	Nakamura et al., 1965a
Mouse	intraperitoneal (in Tween 20)	ca. 5-10	Yoshikawa and Kawai, 1966
Mouse	intraperitoneal (in polyethylene glycol)	ca. 25	Doull et al., 1962
Mouse	intraperitoneal (in polyethylene glycol)	34.5	Yoshikawa and Kawai, 1966
Mouse	intraperitoneal (in polyethylene glycol)	34.5	ISK, 1991
Mouse	intraperitoneal (in olive oil)	36.9	Yoshikawa and Kawai, 1966
Mouse	intraperitoneal (in olive oil)	36.9	ISK, 1991
Mouse	intraperitoneal (in gum arabic)	47	Cheav and Foussard-Blanpin, 1990
Mouse	intraperitoneal (in tragacanth emulsion)	50	BASF, 1957
Mouse	intraperitoneal (aqueous suspension with tragacanth)	ca. 50	BASF, 1969a, b; Zeller et al., 1969
Mouse	intraperitoneal	50 tail-lifting reaction after 20 to 30 minutes, repeated convulsions, death after 60 to 90 minutes	Yoshikawa, 1967
Mouse	subcutaneous (in Tween 20)	7.94	Ueda, 1961
Mouse	subcutaneous (in polyethylene glycol)	46.4	Yoshikawa and Kawai, 1966
Mouse	subcutaneous (in polyethylene glycol)	46.4	ISK, 1991
Mouse	subcutaneous (in propylene glycol)	ca. 50	Thiess, 1968

Table 1. (continued)

Species	Route (preparation)	LD$_{50}$ (mg/kg body weight) or dose at which effects seen	Reference
Mouse	subcutaneous (tragacanth emulsion)	55	BASF, 1957
Rat	oral (in "guar gum")	30 (males) 71 (females)	Eastman Kodak, 1992
Rat	oral (in propylene glycol)	ca. 35	Thiess, 1968
Rat	oral (in gum arabic)	100	Cheav and Foussard-Blanpin, 1990
Rat	oral (aqueous suspension with tragacanth)	ca. 125	BASF, 1969a, b; Zeller et al., 1969
Rat	oral (tragacanth emulsion)	150	BASF, 1957
Rat	intraperitoneal (in gum arabic)	50	Cheav and Foussard-Blanpin, 1990
Rat	intraperitoneal (in gum arabic)	62	Nakamura et al., 1965a
Rat	dermal (aqueous suspension, 4 hours/10% of the surface of the body, abdominal skin)	convulsions, dose unspecified	Thiess, 1968
Rabbit	oral (in propylene glycol)	12.5 survived 25 convulsions, 50 lethal	Thiess, 1968
Rabbit	oral (aqueous suspension with tragacanth)	12.6 survived, weight loss 25 weight loss, lethal after 4 days 50 lethal after 2 days 70 lethal within 24 hours	BASF, 1958
Rabbit	oral (aqueous suspension)	25 convulsions, lethal	Zeller et al., 1969
Rabbit	oral (in the diet)	50 tolerated 100 convulsions, lethal after 22 hours	Thiess, 1968
Rabbit	intraperitoneal (in Tween 20)	10-20 some convulsions 25 convulsions	Ito, 1971, 1972
Rabbit	subcutaneous (no data)	10 tolerated 50 convulsions, survived 100 convulsions, lethal after 22 hours	Thiess, 1968
Rabbit	dermal (50% aqueous preparation, 20 hours)	up to 5000 tolerated	Thiess, 1968; Zeller et al., 1969
Rabbit	dermal (10-50% preparation in distilled water, propylene glycol or peanut oil, 20 hours)	up to 5000 tolerated	Kleinsorge et aL., 1979

Table 1. (continued)

Species	Route (preparation)	LD$_{50}$ (mg/kg body weight) or dose at which effects seen		Reference
Cat	oral (aqueous suspension with tragacanth)	25 survived, weight loss, shortness of breath, salivation 50 lethal within 1 to 2 hours 70 lethal within 1 to 2 hours		BASF, 1958
Rat	inhalation	with air enriched with volatile components at 20 °C, 8 hours	without effects	BASF, 1969a, b; Zeller et al., 1969
Rat Mouse	inhalation dynamic	fine dust, only just perceptile to the eye, for 8 hours, concentration not determined	convulsions, death after 24 hours	Thiess, 1968
Mouse Rat Guinea pig Rabbit	inhalation static	sublimation at 170 to 180 °C, concentration 3-5 mg/l (calculated)	mouse died, augmented reflexes in the rat, guinea pig survived, rabbit survived	Thiess, 1968

independent of the various routes of administration, disturbances of balance, augmented reflexes and tonic-clonic convulsions occurred in all species following absorption of o-phthalodinitrile.

Subacute toxicity

Oral administration

In a preliminary study, the oral administration of 5 mg/kg body weight twice weekly was tolerated 26 times by a cat, which vomited and suffered disturbances of balance. The animal recovered after the treatment was discontinued (Kleinsorge et al., 1979).

Oral administration of 5 mg/kg body weight twice weekly for 3 months (25 doses) was tolerated by a cat, which suffered retching and disturbances of balance, while single oral doses of 20 and 50 mg/kg rapidly led to death (Nakamura et al., 1965a).

Dermal application

Rats which had ca. 10% of their body surface area exposed to a suspension of o-phthalodinitrile for 4 hours on each of 3 consecutive days suffered convulsions and some of the animals died (no further details; Thiess, 1968).

Weight loss and over-excitation occurred after 1 to 2 applications of a 50% suspension of o-phthalodinitrile in a sweat simulant to the clipped abdominal skin of rats (duration of exposure 4 hours in each case; no further details; Kleinsorge et al., 1979).

2 groups of rats (weight ca. 80 g, strain and sex unspecified) were given subcutaneous injections of 10 µg o-phthalodinitrile or had 2 mg of a mixture of o-phthalodinitrile in hydrophillic ointment rubbed into their necks; in both cases the treatment was carried out daily for 30 days. On day 32, blood was taken from the rats in the treated groups and from a control group. There was a significant reduction in plasma protein and in the erythrocyte count and a significant increase in the leukocyte count and in thymol clouding in the treated rats compared with the control group. No significant differences were found between the two treated groups. The author concluded that o-phthalodinitrile can cause functional disturbances of the liver and anaemia and that absorption through the skin is possible (no further details; Ueda, 1961).

A cat that was treated 20 times with 50 mg o-phthalodinitrile/kg body weight which was applied to the skin of the neck on 5 days/week showed no signs of toxicity, although the repeated application of 100 mg/kg body weight subsequent to this caused a convulsive attack after which an unsteady gait was evident. However, at the end of the study, the cat recovered (Thiess, 1968).

Cats tolerated the application of a single dermal dose of 50 mg/kg body weight for 8 hours and 40 dermal exposures to 50 or 100 mg/kg body weight (no further details), with only one convulsive attack (with subsequent unsteady gait) occurring after the repeated application of 100 mg/kg body weight (Kleinsorge et al., 1979).

The dermal application of 50 mg/kg body weight 20 times with subsequent dermal application of 100 mg/kg on 21 occasions caused no signs of toxicity in cats (Zeller et al., 1969).

Thus on dermal application to cats, signs of systemic toxicity were observed only after repeated exposure to large doses (from ca. 100 mg/kg body weight).

7.3   Skin and mucous membrane effects

The skin irritancy of o-phthalodinitrile was studied in a "patch test" by applying a cotton patch (ca. 2.5 cm x 2.5 cm) coated with the substance to the clipped dorsal skin of white rabbits for 15 minutes or 20 hours. After the 15-minute exposure, the treated skin was washed off, first with neat polyethylene glycol 400 and then with a 50% aqueous polyethylene glycol 400 solution; after the 20-hour exposure, the skin was not further treated. The effects were recorded on removal of the dressing and after 1, 3 and 8 days. The substance was not irritating to the skin after exposures of either 15 minutes or 20 hours (Zeller et al., 1969).

When o-phthalodinitrile was applied to the skin of the rabbit ear on a wad of cotton wool for 24 hours either in concentrated form, as a 10% solution in solvent (no further details) or as a 10% solution in oil, mild, transient redness was observed with increased blood supply to the treated ear, and the substance was thus evaluated as not irritating (no further details; Thiess, 1968).

In rabbits, the application of o-phthalodinitrile (ca. 100% or 98.7% pure) as a 50 % aqueous preparation to the dorsal skin for 1, 5 or 15 minutes or 20 hours or to the skin of the ear for 20 hours (no further details) questionably gave rise to slight reddening of the back and caused slight reddening of the ear after 20 hours, effects which were completely reversible after 8 days. The substance was thus evaluated as not irritating to the skin of the rabbit (BASF, 1969a, b).

When 0.25 to 1.0 g o-phthalodinitrile/kg body weight was applied as an aqueous preparation to the skin of 3 guinea pigs and covered with an impermeable cuff for 24 hours, the treated skin showed mild to moderate oedema after 24 hours, sloughing and mild alopecia after a week and sloughing after 2 weeks. In this study, the authors evaluated o-phthalodinitrile as a mild skin irritant without indications of systemic effects (Eastman Kodak, 1992).

To study the acute eye irritancy, ca. 50 mg o-phthalodinitrile were introduced into the conjunctival sac of the rabbit eye (number of animals not specified) and the effects were determined after 10 minutes, 1 and 24 hours and 3 and 8 days. The substance caused slight congestion of the vessels (Zeller et al., 1969) which was evidently transitory, so that in accordance with current criteria, the substance would be evaluated as not irritating to the eye of the rabbit (BASF, 1993; Kleinsorge et al., 1979).

When 50 mm^3 o-phthalodinitrile (ca. 100% or 98.7% pure) were introduced into the rabbit eye (no further details), marked reddening and marked oedema were found after 1 hour, with slight redness after 24 hours, effects which were completely reversible after 8 days. In the control eye, which had been treated with talcum, the same effects were evident after 1 hour with even more marked effects after 24 hours. o-Phthalodinitrile was evaluated as not irritating to the rabbit eye (BASF, 1969a, b).

See also Sect. 8 on experience in humans (Thiess, 1968).

7.4   Sensitisation

No information available.

7.5   Subchronic and chronic toxicity

See Sect. 7.10 (in particular Bio Research, 1995b).

7.6   Genotoxicity

7.6.1 In vitro

o-Phthalodinitrile (97% pure) was studied in the Salmonella/microsome test in strains TA 98, TA 100, TA 1535 and TA 1537 at concentrations of 3.15 to 3000 µg/plate with metabolic activation (S9-mix from Aroclor 1254-induced rat liver) and 31.5 to 3000 µg/plate without metabolic activation. The concentrations used were neither cytotoxic nor mutagenic. There was also no mutagenic activity in strain TA 98 compared with positive and negative control substances when 1,1,1-trichloropropene-2,3-oxide (TCPO), which inhibits epoxide hydratase and reduces glutathione, was added (Oesch, 1978).

In an HPRT test in Chinese hamster V79 cells which was carried out in accordance with OECD guidelines, the point-mutagenic activity of o-phthalodinitrile (98.8% pure) dissolved in DMSO was studied at concentrations of 23.0, 90.0, 160.0 and 230.0 µg/ml for 4 hours with and without S9-mix. At the highest concentration, the cell survival rates were 86.9 and 90.9% without S9-mix and 92.0 and 80.6% with S9-mix (2 independent experiments). o-Phthalodinitrile did not show any mutagenic activity in either experiment (LMP, 1987a).

7.6.2 In vivo

In a micronucleus test in mice which was carried out in accordance with OECD guidelines, no increases in the incidence of micronuclei were found after 6, 24 or 48 hours following the administration of 2 doses of o-phthalodinitrile (ca. 99 % pure, dissolved in dimethyl sulfoxide) by gavage at an interval of 24 hours (doses tested 2, 7 and 20 mg/kg body weight; examination of 1000 bone marrow cells in each case in groups of 10 mice (5 males, 5 females); LMP, 1987b).

See also Sect. 8 (Fleig and Thiess, 1979).

7.7   Carcinogenicity

The studies presented in Table 2 were carried out to investigate the carcinogenic activity of o-phthalodinitrile:

Table 2. Protocol of long-term (2 year) studies in mice and rats

	Numbers of animals[1]	Type of administration	Dose
Mouse[2]	50	oral (diet) 5/week	2.0, later 1.0 mg/mouse (1 mg/mouse $\cong$ ca. $^1/_6$ $LD_{50)}$
	51	subcutaneous 1/10 days	2.0, later 1.0, 0.5, 0.25, 0.2 mg/mouse (0.2 mg/mouse $\cong$ ca. $^1/_{30}$ LD_{50})
	48	dermal 2/week	2.0, later 0.8 mg/mouse (0.8 mg/mouse $\cong$ ca. $^1/_8$ LD_{50})
Rat[3]	47	oral (diet) 5/week	5 mg/rat
	50	subcutaneous 1/10 days	5 mg/rat

[1] Sex distribution given only roughly as "in approximately the same proportions"
[2] CC-57-W strain
[3] Rapollo strain, sterile

The males and females (no precise details of sex distribution) were ca. 3 months old at the beginning of the study and the duration of exposure was 2 years. Because of cumulative toxicity in the mice (no further details) the dose was gradually reduced within the first 5 months (see Table 2). The last doses specified were well tolerated and were administered for the remaining $1^{1}/_2$ years. For each type of administration, o-phthalodinitrile was initially suspended in sunflower oil. Calculated in terms of body weight, the mice (assumed weight 40 g) initially received ca. 50 and later 25 mg/kg body weight/day in the diet (average total dose ca. 427 mg), ca. 50 and then 20 mg/kg body weight/day dermally (average total dose ca. 141 mg) and ca. 50 and later 5 mg/kg body weight/day subcutaneously (average total dose ca. 16 mg). The corresponding figures for the rats (assumed weight 400 g) were ca. 12.5 mg/kg body weight/day in the diet (average total dose ca. 2010 mg) and ca. 12.5 mg/kg body weight/day subcutaneously (average total dose ca. 342 mg). The latency period and location of the tumours are shown in Table 3.

In the animals with leukaemia, perivascular leukaemic infiltrates were found in the lymph nodes, kidneys, liver and spleen in the mice. Poorly differentiated lymphoma-like leukoses as well as lymphatic and myeloid leukoses were described. Similar histological changes also occurred in the rats. Details of the tumour incidences in control groups of mice and rats and historical control data were not presented. It was only generally reported that leukaemia was observed in 6.8% of cases in control mice (no absolute figures), that is 10 to 12 times less than in these experiments (Pliss and Wolfson, 1972; Pliss, 1972).

This study can only be evaluated to a limited extent as it has been inadequately described. Information on concomitant control groups and the causes of death in the numerous mice and rats which were not evaluated is not provided and the documentation of individual findings is insufficient. Norpoth (1983) suggested that the initial administration of toxic doses with cumulative effects could possibly have led to activation of latent leukaemia viruses in mice, accounting for the increased number of leukoses.

Table 3. Time of tumour manifestation and distribution of tumours in mice and rats according to route of exposure

Species	Route	Number of animals evaluated		Day of manifestation of first tumour	Distribution of tumours							
		total	number with tumours		Leukosis	Thyroid gland	Mammary gland	Ovary Prostate	Gut	Liver	Mesen-terium	Local
Mouse	oral	23	23	457	23					1[a]		
	dermal	27	20	324	19			1[b]				
	s.c	22	20	443	20							1[c]
Rat	oral	18	8	461	6	1	1[d]	1[e]		2[f]	1[h]	
	s.c	28	22	145	15		3[d]		1[i]	4[g]		1[c]

s.c. Subcutaneous

a	Haemangioma	f	Hepatosarcoma
b	Ovarian folliculoma	g	1 Cholangioma, 3 hepatosarcomas
c	Sarcoma	h	Lymphangioma
d	Fibroadenoma	i	Adenocarcinoma
e	Prostate adenoma		

7.8   Reproductive toxicity

No information available.

7.9   Effects on the immune system

No information available.

7.10 Neurotoxicity

On administration of toxic doses of o-phthalodinitrile to mice (female, weight 20 to 25 g, strain ICR) in an acute toxicity test, a spontaneous tail-lifting reaction (Straub phenomenon) occurred after 5 to 30 minutes, followed by generalised convulsive attacks lasting 5 to 10 seconds and vocalisation (squealing). The tail-lifting reaction and convulsive attacks were repeated several times at intervals of 30 to 60 seconds and death occurred within 5 hours (see also Sect. 7.2; Yoshikawa and Kawai, 1966). A study of the effect of anticonvulsants on the neurotoxic effect of o-phthalodinitrile revealed that acetylcholine, methacholine and acetazolamide lessened the effects of o-phthalodinitrile and markedly reduced the death rate (no further details), while pyridoxine, also an anticonvulsant, showed no effects of this type. The authors compared these results with the convulsive effect of n-butylpyrrolidine and semicarbazide which, in contrast to those of o-phthalodinitrile, were decreased by pyridoxine, while acetylcholine and acetazolamide caused no decrease in convulsions in this case (Yoshikawa, 1967).

Groups of 2 to 3 white Wistar rats (weight ca. 150 g) were given single doses of 10, 30, 50, 70, 100, 150 or 300 mg o-phthalodinitrile/kg body weight by intraperitoneal injection or 1000 mg/kg body weight orally. The brains of the rats were examined after their deaths or 7 days after administration of the substance, with the aid of various histopathological staining methods. Degenerative effects on the nerve fibres in the hypothalamus (corpus luysi) and the formatio reticularis of the mesencephalon were found in both high dose groups only with the Nauta-Gygax staining method (silver impregnation) which was used in only one rat in the 10 mg/kg group, 2 in the 30 mg/kg group, and one rat in each of the 50 and 70 mg/kg groups (Nakamura et al., 1965a).

Groups of 4 white Wistar rats (weight ca. 150 g) were given intraperitoneal injections of 0.01, 0.1, 1 or 10 mg o-phthalodinitrile/kg body weight/day for 32 days (one group for 33 days), of 1 or 10 mg/kg body weight/day for 83 days or of 0.01 or 0.1 mg/kg body weight/day for 90 days (equivalent to $^1/_{6000}$ to $^1/_6$ of the LD_{50}, dissolved in gum arabic). The brain tissues of the rats were then examined and the results compared with those from 5 untreated controls which were similarly examined after 83 (one animal) or 90 days. Macroscopically, there was no damage to the meninges or the surface of the brain and no haemorrhages, although hypertrophy of the right cerebral ventricle occurred in one rat which had received 10 mg/kg body weight/day for 32 days. Histomorphological studies with the aid of various staining techniques revealed no pathological effects in most cases. Only with Nauta-Gygax staining (silver impregnation) could nerve fibre anomalies (in the hypothalamus, mid-brain, reticular nerve tissue of the pons and in the thalamic nuclei, particularly the lateral and posterior nuclei) be detected in 2 rats in the high dose group (10 mg/kg body weight/day for 32 or 83 days). These degenerative nerve fibre anomalies were shown by Nissl's staining to include chromatolysis, pyknosis of the cytoplasm and cell nucleus, vacuolisation and homogenisation and destruction of the cells in the hypothalamus and particularly in the underlying thalamic nucleus (corpus luysi). These effects were also seen to a lesser extent in 2 rats that had received 1 mg/kg body weight/day for 32 or 83 days. These degenerative effects on the nerve fibres were compared with the same, if less marked, results after a single dose by intraperitoneal injection (see Nakamura et al., 1965a). Through these studies, the primary lesions could be localised in the hypothalamus and the corpus luysi (Nakamura et al., 1965b).

EEG studies are said to have shown that o-phthalodinitrile inhibits the excitation in the thalamus and mid-brain caused by electrical stimulation. After administration of o-phthalodinitrile for 60 days, histological examination apparently revealed mild phagocyte infiltration and atrophy of the brain cells (no information provided on dose or type of atrophy observed; Ito, 1971, 1972).

In a dose-finding study prior to a 90-day neurotoxicity study, groups of 10 male and 10 female Sprague-Dawley rats (ca. 50 days old, weight 243 to 288 g and 161 to 196 g, respectively) were given o-phthalodinitrile (99.3% pure) at 5, 15 or 45 mg/kg body weight/day (target concentrations; measured doses ca. 4.8, 14 and 34 mg/kg body weight) in the diet for 14 days. To establish the possible neurotoxic effects of the substance, all of the rats were studied in a "Functional Observational Battery" (FOB; in accordance with US EPA guidelines) before and on days 1, 7 and 14 of dosing. The study was carried out in accordance with OECD guideline no. 407. The FOB included the observation and recording of numerous qualitative parameters as well as quantitative measurement of fore- and hind-limb grip strength, body temperature and hind-limb splay after falling from a height of 30 cm. In the highest dose group and the control group, 3 rats/sex/group underwent perfusion to investigate neuropathology. The rats in the highest dose group (45 mg/kg/day) showed a significantly lower body weight compared with the controls (p<0.01) on days 1, 7 and 14, on which body weight was recorded, while in the intermediate dose group, body weight was only reduced in the males. Neither the neurotoxicological nor the haematological or clinical chemistry studies revealed any toxicologically relevant significant differences compared with the control group. A few organ weight changes were the result of the marked reduction in body weight, as no histopathological or biochemical correlates were found. Nothing abnormal was detected in the histopathological or neuropathological studies (Bio Research, 1995a).

Based on the results of the 14-day dose-finding study described above (Bio Research, 1995a), the 90-day oral neurotoxicity study with 99.3% pure o-phthalodinitrile was conducted with doses of 3, 8 and 25 mg/kg body weight/day for male rats, which were somewhat more susceptible, and 3, 10 and 30 mg/kg body weight/day for the females. The study was carried out according to OECD guideline no. 408 and the neurotoxicity studies in accordance with EPA guidelines; the substance was added to the diet. In each case 5 female and 5 male Sprague-Dawley rats from the total 15 rats/sex/ group underwent an interim post-mortem after 28 days to study possible cumulative or delayed effects. The remaining 10 animals/sex/group were studied for neurotoxic effects in a Functional

Observational Battery (FOB) before the beginning of the study and in weeks 3, 7 and 12. At the end of the study, 5 of the 10 animals/sex/group underwent macroscopic and microscopic examination, while the remaining 5 animals/sex/group underwent perfusion to allow a thorough neuromorphological examination. This was carried out in the high dose group and the results were compared with those from the untreated controls. The results of the study showed that during the entire 3-month period, there was a significant ($p<0.01$) reduction in body weight compared with the controls in males and females in the high dose groups (25 and 30 mg/kg, respectively) and a reduction in body weight in the females in the intermediate dose group (10 mg/kg) which was not significant, effects that were correlated with the significantly reduced feed intake. Significant effects ($p<0.05$) in males and females in the high dose group during the 3 FOB studies included an increase in the level of excitation (hyperactivity) and increased vocalisation on transfer from the cage to the test arena as well as, in males, an increase in rearing in the arena. Motor activity was measured quantitatively before the beginning of the study and in weeks 3, 7 and 12, six times at 10 minute intervals over one hour. In each case activity was determined in figure 8 mazes with the aid of light beams. The following significant results were obtained: In males, the activity in the high dose group at the end of the study (week 12) was significantly increased ($p<0.001$; total number of light signals), with a clear significant linear dose-response ($p<0.001$) for the total number of light signals and the linear-constructed variable. The linear-constructed variable was reduced within the 1-hour test phase from the first to the sixth 10-minute test period, in a significant dose-dependent manner compared with the control, that is the normal reduction in activity levels from the first to the sixth 10-minute period during the 1-hour observation phase did not occur (equivalent to a high linear-constructed variable and high rate of linear change from the first to the sixth period within the 1-hour observation phase), but the activity level remained elevated, in a dose-dependent manner, during the 1-hour observation phase from the first to the sixth period (corresponding to low linear-constructed variable from the first to the sixth period and a low linear rate of change. In the females, the linear-constructed variable was significantly reduced (p<0.01 or p<0.001) as early as week 3 and was also reduced in weeks 7 and 12 in the intermediate and high dose groups (10 and 30 mg/kg/day) in a significant dose-dependent manner; the rate of linear change was also reduced in the 1-hour observation phase (only in week 3). In weeks 7 and 12, the activity (total number of light signals) was also increased in the females in the intermediate and high dose groups in a significant dose-dependent manner. The substance thus caused a significant increase in the rate of activity in a dose- and time-dependent manner in the females in the intermediate and high dose groups and in the males in the high dose group only, as early as week 3 in the females, but only from week 12 in the males. Eye examinations revealed no effects after 4 weeks in the interim group, while at the end of the study, several male and female rats in the high dose group (25 and 30 mg) and some females in the intermediate dose group (10 mg) showed effects on the eyes in the form of localised posterior clouding of the lens. Apart from significantly increased phosphorous levels (p<0.05) in both sexes in the high dose group and increased alkaline phosphatase, potassium and cholesterol levels in females in the high dose group at the end of the study, laboratory studies showed no toxicologically relevant haematological or biochemical effects after 4 weeks or 3 months. There were also no toxicologically relevant changes in organ weights or macromorphological effects after 4 weeks or 3 months in the treated groups compared with the controls. The microscopic examination of the testes and epididymides showed no histopathological effects. Histopathological examination revealed hyaline droplets in the kidney tubules in all males at the top dose level and in most males at the intermediate dose level after 4 weeks and 3 months. These effects are known to occur specifically in male rats and have no toxicological relevance to humans. No treatment-related effects were evident in the neuropathological studies involving specific neurohistopathological staining methods in the different regions of the central and peripheral nervous system. The main subchronic effects of o-phthalodinitrile were thus reduced growth accompanying reduced feed intake, increased levels of activity and clouding of the lens in both sexes at the top dose level and to some extent in females at the intermediate dose level. The no observed adverse effect level was therefore 3 mg/kg body weight/day (Bio Research, 1995b).

7.11 Other effects

To investigate the possible radio-protective effect of o-phthalodinitrile, groups of 10 male mice (6 to 8 weeks old, CF_1 strain, weight 20 to 25 g) were given intraperitoneal injections of 5 or 10 mg o-phthalodinitrile/kg body weight and were then irradiated (whole body) with 800 r X-rays 10 to 15 minutes later. In order to be able to use the maximum tolerated dose, the approximate LD_{50} after 7 days (ca. 25 mg/kg body weight) was determined beforehand. The control group, which was treated with the vehicle only prior to irradiation, and the treated group were both observed for 30 days. Irradiation led to the deaths of all of the animals used within 2 weeks, including the control animals. The average survival time of the control mice was 9±3 days. A radio-protective effect would have been assumed if the average survival time were increased by ≥ 5 days. o-Phthalodinitrile decreased the average survival time by 4 days in the high dose group and by 2 days in the low dose group, and thus no radio-protective effect was established (Doull et al., 1962).

8. Experience in humans

Of 20 workers employed in an o-phthalodinitrile production plant between July 1957 and March 1959, 10 showed signs of toxicity within this period, including sudden disturbances of consciousness with clonic spasms and bradycardia, sometimes with retrograde amnesia. These attacks occurred without preliminary symptoms in the midst of activity (such as cycling, mountaineering; no detailed data on time between attacks and last exposure) and went away without noticeable after-effects. Between the beginning of 1959 and June 1959, 3 cases of jaundice also occurred in the same production plant, which Ueda was commissioned to investigate in July 1959 by carrying out blood analyses in 21 employees. The average age was 26.6 (19 to 43) years, the duration of employment ranged from a few months to $4^1/_2$ years (average 18 months). With increasing duration of employment, a reduction in the haemoglobin content and the erythrocyte count occurred compared with normal levels (no further details on control groups). The average body weight of the 21 exposed subjects examined (52.9 kg) was below the average body weight for 26- to 29-year olds in the year 1957 (55.68 kg). The author concluded that signs of central nervous system toxicity together with generalised convulsions and damage to the blood can occur on exposure to large amounts of o-phthalodinitrile (Ueda, 1961). No data are available on exposure levels, but when the attacks described occurred, the workers were evidently exposed to the substance without protection, as it is reported that the wearing of masks, hand washing and gargling were only ordered because of the incidents, and no further cases of toxicity were apparently observed thereafter.

Acute intoxication has been described after absorption through the skin and following inhalation of the dust during the manufacture of o-phthalodinitrile: heavy perspiration and prolonged skin contact favoured percutaneous absorption. The signs of toxicity occurred after an average latent period of about 12 hours ($^1/_2$ to 48 hours) and included dizziness, nausea, retching, vomiting, headaches, sudden loss of consciousness (sometimes for short periods or for over an hour), and epileptiform convulsions (tonic-clonic, generalised with foaming at the mouth, lasting 1 to 3 minutes, sometimes with retrograde amnesia, without subsequent positive findings on examination, including by EEG). Irritation of the skin and mucous membranes was observed (Thiess, 1968; Zeller et al., 1969; Thiess and Kleinsorge, 1977).

No abnormal findings were evident on questioning, on clinical and neurological examination or in laboratory studies of clinical chemistry or haematological parameters (to determine effects on the blood, liver and urine) in 81 workers who were occupationally exposed to o-phthalodinitrile (average duration of exposure 8.5 years; range 1 to 34 years), 11 of whom had suffered acute intoxication with o-phthalodinitrile; no effects indicative of illness were detectable by EEG in 15 subjects who were particularly exposed (11 of whom had previously experienced acute intoxication; Kleinsorge et al., 1979).

Some time after o-phthalodinitrile intoxication due to dust exposure, a worker suffered retching, vomiting and dizziness, a generalised convulsive attack and then a fracture of the base of the skull with fatal results. Subsequent examination of the brain revealed no histopathological effects typical of a convulsive condition, although epi- and subdural haematomas, a cortical focus of contusion and brain oedema attributable to the fractured skull were evident (Thiess, 1968).

In an epidemiological mortality study, a follow-up of 83 retired workers was carried out, each of whom had been exposed to o-phthalodinitrile for longer than 6 months (average duration of monitoring 9 years, exposures of up to 24 years). The 13 deaths observed in this cohort, when compared with the number of deaths to be expected in the Federal Republic of Germany (11.92), was still within the range of the statistical limits of variation. The total of 4 observed malignancies (2 lung carcinomas, one stomach carcinoma, one acute myeloid leukaemia) was not increased compared with the expected number (Frentzel-Beyme et al., 1979).

Studies of the chromosomes were carried out in lymphocytes cultured from 20 industrial workers (age 32 to 62 years, average 42.6 years) with possible o-phthalodinitrile exposures ranging from 2 to 24 years (average 12.8 years) and the results were compared with those from a corresponding control group. Of the 20 workers studied, 7 had experienced acute intoxication with o-phthalodinitrile during their employment (as described in Thiess, 1969). The workers were questioned about their alcohol consumption, nicotine and medicine use and viral infections suffered, and all findings were compared with those from a control group of 20 people who had not been exposed to chemicals. Lymphocyte cultures were incubated at 37 °C for 70 to 72 hours and 100 metaphases were examined for chromosome aberrations in each case. Neither the group as a whole nor the 7 people within the group who had experienced acute intoxication with o-phthalodinitrile showed significant differences compared with the control group. The chromosome exchange rate was increased, but the authors did not conclude that the substance had mutagenic activity from this as exchanges were also observed in the non-exposed control groups (Fleig and Thiess, 1979).

In 1936, 6 people with healthy skin had solutions of o-phthalodinitrile (1% in oil and 10% in an undefined solvent) on an occlusively-covered cotton wool pad applied to $1\frac{1}{2}$ cm^2 sites on the inside of the forearm for 24 hours. While the 1% solution did not cause any irritation, the 10% solution in all cases gave rise to mild reddening, which temporarily became more marked in 4 subjects during the course of the following hours. No signs of systemic toxicity were observed in any of the volunteers. It was reported that one worker reacted with an allergic-toxic dermatitis on contact with o-phthalodinitrile dust (Thiess, 1968).

Between 1 January 1989 and 30 April 1995, no effects related to o-phthalodinitrile were noticed in the out patients section of an occupational medicine department. In occupational medical examinations of employees of 1 production and 3 processing plants, between 1 January 1989 and 31 December 1994, no differences concerning affects on the airways and skin were seen compared to the rest of the company (BASF, 1995). Therefore, the substance should not be considered as having a strong sensitising potential.

9.   Threshold limit values

No information available.

References

BASF AG, Gewerbehygienisch-Pharmakologisches Institut
Vergleichende Untersuchungen über die Toxizität von Ortho- und Isophthalodinitril an kleinen Nagetieren
Unpublished report no. VII/178 (1957)

BASF AG, Gewerbehygienisch-Pharmakologisches Institut
Vergleichende Untersuchungen über die Toxizität von Ortho- und Isophthalodinitril an höheren Tieren
Unpublished report no. VII/178, -262, VII/295 (1958)

BASF AG, Gewerbehygienisch-Pharmakologisches Institut
Gewerbetoxikologische Vorprüfung, o-Phthalodinitril, ca. 100 %ig
Unpublished report no. XVIII307 (1969a)

BASF AG, Gewerbehygienisch-Pharmakologisches Institut
Gewerbetoxikologische Vorprüfung, o-Phthalodinitril, ca. 98,7 %ig
Unpublished report no. XVIII308 (1969b)

BASF AG
AIDA-Grunddatensatz 1,2-Benzenedicarbonitrile (9CI) (1993)

BASF AG, Abteilung Arbeitsmedizin und Gesundheitsschutz
Written communication to BG Chemie, 09.05.1995

Bio Research Laboratories Ltd.
A 2-week dietary toxicity, behavioral and neuromorphological
study of o-phthalodinitrile in the rat
Unpublished report , project no. 97206 (1995a)
On behalf of BG Chemie

Bio Research Laboratories Ltd.
A 13-week dietary toxicity, behavioral and neuromorphological
study of o-phthalodinitrile in the rat
Unpublished report, project no. 83536 (1995b)
On behalf of BG Chemie

Cheav, S.L., Foussard-Blanpin, O.
Etude comparée de la toxicité de dérivés cyanés et/ou amides arom-
atiques
Ann. Pharm. Fr., 48 (1), 23–31 (1990)

Doull, J., Plzak, V., Brois, S.J.
A survey of compounds for radiation protection
USAF Radiation Laboratory, University of Chicago, Chicago, Illi-
nois, pp. 1–3, 68 (1962)
National Technical Information Service, U.S. Department of Com-
merce, Springfield, VA 22161, AD-277689, Nov. 86

Eastman Kodak Company
1,2-Benzenedicarbonitrile: acute oral toxicity test in rats and skin
irritation test in guinea pigs with cover letter dated 073092 (1992)
NTIS/OTS0540933

Fleig, I., Thiess, A.M.
Chromosomenuntersuchungen bei Mitarbeitern mit Exposition gegenüber ortho-Phthalodinitril (o-PDN)
Zentralbl. Arbeitsmed., 29, 127–129 (1979)

Frentzel-Beyme, R., Thiess, A.M., Wieland, R.
Mortalitätssurvey bei Mitarbeitern aus der ortho-Phthalodinitril-Produktion
Zentralbl. Arbeitsmed., 29, 121–127 (1979)

ISK Biotech Corporation
Letter from ISK Biotech Corporation to USEPA submitting enclosed material safety data sheets, health & safety studies & information concerning 1,3-dicyanobenzene with attachments (1991)
NTIS/OTS0533488

Ito, J.
The experimental study on the neurotoxic action of phthalodinitrile (English abstract)
Kansai Ika Daigaku Zasshi, 23, 93–126 (1971)
Also in: Chem. Abstr., 77, 71015j (1972)

Kleinsorge, H., Thiess, A.M., Zeller, H.
Untersuchungen zur Morbidität bei Mitarbeitern aus der ortho-Phthalodinitril-Produktion
Zentralbl. Arbeitsmed., 29, 130–132 (1979)

LMP (Laboratorium für Mutagenitätsprüfung, Technische Hochschule Darmstadt)
ortho-Phthalodintrile, OPDN – detection of gene mutations in somatic mammalian cells in culture: HGPRT-test with V79 cells
Unpublished report LMP 271 A (1987a)
On behalf of BG Chemie

LMP (Laboratorium für Mutagenitätsprüfung, Technische Hochschule Darmstadt)
ortho-Phthalodinitril – micronucleus test in bone marrow cells of the mouse
Unpublished report LMP 271 B (1987b)
On behalf of BG Chemie

Matsushita, H.
Umgang mit chemischen Substanzen und Gesundheitsüber-
wachung. (37). Vergiftung durch o-Phthalodinitril und Gesundheits-
überwachung (German translation of the Japanese)
Sangyo Igaku Janaru, 4 (4), 10–15 (1981)

Nakamura, K., Ohyanagi, H., Suzuki, A.
Histological studies on the rat brain in case of acute phthalodinitrile
intoxication
Kobe J. Med. Sci., 11, 63–72 (1965a)

Nakamura, K., Ohyanagi, H., Suzuki, A.
Histological studies on the rat brain in case of chronic phthalodini-
trile intoxication
Kobe J. Med. Sci., 11, 139–150 (1965b)

Norpoth, K.
Phthalic anhydride and some derivatives
In: Parmeggiani, L. (ed.)
Encyclopaedia of occupational health and safety
3rd ed., vol. 1, pp. 1693–1694
International Labour Office, Geneva (1983)

Oesch, F.
Ames test for o-Phthalodinitril
Unpublished report (1978)
On behalf of BASF AG

Pliss, G.B.
Carcinogenic activity of some chemicals (English abstract of a Russian report)
Nekot Itogi Izuch Zagryazneniya Vnesh Sredy Kantserogen Vesh-
chestvami, 103–106 (1972)

Pliss, G.B., Wolfson, N.I.
Über die leukosogene Wirkung des Phthalonitrils (German transla-
tion of the Russian)
Vop. Onkol., 18, 81–86 (1972)

Thiess, A.M.
Beobachtungen von Gesundheitsschädigungen und Vergiftungen durch Einwirkung von o-Phthalodinitril
Zentralbl. Arbeitsmed., 18 (10), 303–312 (1968)

Thiess, A.M.
Beobachtungen von Gesundheitsschädigungen und Vergiftungen durch Einwirkung von o-Phthalodinitril
Zentralbl. Arbeitsmed. Geschwulstforsch., 19, 2–15 (1969)
Cited in: Fleig and Thiess (1979)

Thiess, A.M., Kleinsorge, H.
Neurotoxisch wirkende Substanzen und Unfallgeschehen in der chemischen Industrie
Zentralbl. Arbeitsmed., 27 (4), 77–80 (1977)

Towae, F.K., Enke, W., Jäckh, R., Bhargava, N.
Phthalic acid and derivatives
In: Ullmann's encyclopedia of industrial chemistry
5th ed., vol. A20, p. 191–193
VCH Verlagsgesellschaft, Weinheim (1992)

Ueda, H.
Yokohama Igaku, 11 (5), 1425–1456 (1961)
Cited in: Matsushita (1981)

Weast, R.C. (ed.)
CRC Handbook of chemistry and physics
CRC Press, Cleveland, Ohio (1977/78)

Yoshikawa, H., Kawai, K.
Toxicity of phthalodinitrile and tetrachlorophthalodinitrile
Ind. Health, 4, 11–15 (1966)

Yoshikawa, H.
Beziehung zwischen durch toxische Substanzen verursachten Konvulsionen und Antikonvulsiva (German translation of the Japanese)
Igaku to Seibutsugaku, 75 (4), 131–133 (1967)

Zeller, H., Hofmann, H.T., Thiess, A.M., Hey, W.
Zur Toxizität der Nitrile (Tierexperimentelle Untersuchungsergeb-
nisse und werksärztliche Erfahrungen in 15 Jahren)
Zentralbl. Arbeitsmed., 19 (8), 225–238 (1969)

4-Nitro-4'-aminodiphenylamine-2-sulfonic acid

1. Summary and assessment

4-Nitro-4'-aminodiphenylamine-2-sulfonic acid is of low acute toxicity to rats (LD50 oral >5000 mg/kg body weight; LD50 dermal >2000 mg/kg body weight). The substance does not induce methaemoglobin formation in cats at oral doses of 10 and 50 mg/kg body weight.

The subacute oral toxicity of 4-nitro-4'-aminodiphenylamine-2-sulfonic acid is also low. Administration in the diet for 4 weeks is tolerated by rats, up to the highest tested concentration of 12 000 mg/kg feed (1253 mg/kg body weight in males and 1191 mg/kg body weight in females) without toxicologically significant effects.

4-Nitro-4'-aminodiphenylamine-2-sulfonic acid is not irritating to rabbit skin and is not irritating, or at most slightly irritating, to the rabbit eye. It is a clear skin sensitiser in guinea pigs.

The substance is mutagenic in the Salmonella/microsome test both with and without metabolic activation.

An in vitro chromosome aberration test is presently being carried out on behalf of BG Chemie.

2. Name of substance

2.1	Usual name	4-Nitro-4'-aminodiphenyl-amine-2-sulfonic acid
2.2	IUPAC-name	4-Nitro-4'-aminodiphenyl-amine-2-sulfonic acid
2.3	CAS-No.	91–29–2
2.4	EINECS-No.	202–057–9

3.   Synonyms, common and trade names

2-(4-Aminophenylamino)-5-
nitrobenzenesulfonic acid
4-Nitro-4'-aminodiphenyl-
amine-6-sulfonic acid
4-Nitro-4'-aminodiphenyl-
amin-2-sulfonsäure
p-Nitrophenylparaminsäure

4.   Structural and molecular formulae

4.1   Structural formula

O_2N —⬡— $\overset{SO_2H}{\underset{\underset{H}{N}}{}}$ —⬡— NH_2

4.2   Molecular formula $C_{12}H_{11}O_5N_3S$

5.   Physical and chemical properties

5.1   Molecular mass, g/mol   309.30

5.2   Melting point, °C   No information available

5.3   Boiling point, °C   No information available

5.4   Vapour pressure, hPa   No information available

5.5   Density, g/cm³   No information available

5.6   Solubility in water   7.8 g/l (at 20 °C)   (Bayer, 1989)

5.7   Solubility in
organic solvents   No information available

5.8   Solubility in fat   No information available

5.9   pH value   4–5 (at 7.8 g/l water)   (Bayer, 1989)

5.10 Conversion factor   1 ml/m^3(ppm) $\cong$ 12.62 mg/m^3
                                   1 mg/m^3 $\cong$ 0.08 ml/m^3 (ppm)
                                   (at 1013 hPa and 25 °C)

6.   Uses

Intermediate in the production of dyestuffs (Hoechst, 1988).

7.   Experimental results

7.1   Toxicokinetics and metabolism

No information available.

7.2   Acute and subacute toxicity

The acute toxicity was investigated in 5 male (195 to 205 g) and 5
female (185 to 188 g) Wistar rats, which were given the substance by
gavage at a dose of 5000 mg/kg body weight. No deaths occurred
during the 14-day observation period and the LD_{50} value was thus
greater than 5000 mg/kg body weight. Clinical signs of toxicity
observed on the day of administration included motor unrest, ruffled
fur, the adoption of a crouching position, retracted flanks and abnor-
mally open eyes. The urine was stained orange in all of the rats for 3
days after administration of the substance. At post-mortem, reddening of the pancreas was found in some of the males, while nothing
abnormal was detected in the females (Hoechst, 1983a).

A group of 5 male and 5 female SPF Wistar rats (ca. 150 g, about
9 to 14 weeks old) was given a single dose of 5000 mg of the sub-
stance/kg body weight in polyethylene glycol 400 (20 ml/kg body
weight) by gavage. The observation period was 14 days. The treatment was tolerated by all of the rats without signs of toxicity and
there were no effects on body weight gain. The LD_{50} value was thus
greater than 5000 mg/kg body weight (Bayer, 1981).

The acute dermal toxicity of 4-nitro-4'-aminodiphenylamine-2-sulfonic acid was tested in a study based on OECD guideline no. 402. A group of 5 male and 5 female rats (Tif:RAIf, SPF; initial weights 184 to 224 g) had a single dose of 2000 mg/kg body weight of the substance (70.3% pure) applied to their clipped dorsal skin as a 50% preparation in distilled water under semi-occlusive cover. The observation period was 14 days. All of the rats survived. Signs of toxicity observed included dyspnoea, ruffled fur, hunched posture and prostration, as well as temporary sedation and diarrhoea. No local reactions occurred and no abnormal effects were found at post-mortem. The dermal LD_{50} value was thus >2000 mg/kg body weight (Ciba-Geigy, 1984).

4-Nitro-4'-aminodiphenylamine-2-sulfonic acid was given to 2 cats (one per dose, 3.9 and 4 kg, respectively) by gavage at single doses of 10 and 50 mg/kg body weight as a solution in polyethylene glycol 400 (10 and 50 mg/ml, respectively). The methaemoglobin content of the blood was determined (photometrically with the cyan-methaemoglobin method) as was the number of Heinz bodies (nile blue staining, 1000 erythrocytes/smear) before and 3, 7, 24 and 30 hours after treatment. At the 10 mg/kg dose, the methaemoglobin level was 2% at all time points, while at 50 mg/kg it reached a maximum of 3%. The number of Heinz bodies was also not increased, reaching a maximum of 2% (reference values in untreated cats: methaemoglobin level 1.7% (n = 176), Heinz body content 5% (n=163)). Loss of appetite for up to 8 hours was observed after both doses, with general malaise seen in the cat given 10 mg/kg but no effect on general well-being in the cat given the higher dose (Bayer, 1984).

The subacute oral toxicity of 4-nitro-4'-aminodiphenylamine-2-sulfonic acid (58.9% pure, 40.2% water) was studied in male and female Wistar rats (aged about 6 weeks at the beginning of the study) in accordance with OECD guideline no. 407. Groups of 5 males and 5 females were given 0, 480, 2400 or 12,000 mg/kg in the diet for 28 days (equivalent to average substance intakes of 50, 243 and 1253 mg/kg body weight in the males and 47, 234 and 1191 mg/kg body weight in the females). The treatment was tolerated by all of the treated animals, even those given the highest dose, without signs of toxicity. There were no effects on any of the haematological or biochemical parameters studied. Ochre-coloured kidneys were found in 2 males in the 2400 mg/kg group and all males in the 12,000 mg/kg group at post-mortem. Histological examination gave normal findings in all cases. The no effect level was thus (1253 mg/kg body weight in the males and (1191 mg/kg body weight in the females (Hoechst, 1990).

7.3   Skin and mucous membrane effects

In order to study the acute skin irritancy of 4-nitro-4'-aminodiphenylamine-2-sulfonic acid, 3 New Zealand albino rabbits had a patch with 500 mg of the substance (59.8% pure, remainder water) applied to their clipped dorsal skin. This was then covered with a semi-occlusive dressing for 4 hours. The possible irritant effects were evaluated 30 to 60 minutes and 24, 48 and 72 hours after the end of exposure. The test substance was not a skin irritant (Hoechst, 1983b).

In another study, 4-nitro-4'-aminodiphenylamine-2-sulfonic acid (58.1% pure, remainder water) was tested for primary skin irritancy in accordance with OECD guideline no. 404 in 3 adult female albino rabbits (strain HC:NZW, 3.2 to 3.3 kg). A single dose of 500 mg made into a paste with water was applied to the clipped skin of the flank for 4 hours. The findings were evaluated according to the Draize scheme after 1, 24, 48 and 72 hours and after 7 days. No irritant effects were seen at any time (average irritation scores of 0 for erythema and oedema). Thus, 4-nitro-4'-aminodiphenylamine-2-sulfonic acid was not irritating to the skin of the rabbit in this study (Bayer, 1986).

A group of 3 rabbits (albino, New Zealand) each had 100 mg of the test substance (59.8% pure, remainder water) instilled into the conjunctival sac of the left eye. The eye was evaluated 1, 24, 48 and 72 hours after instillation of the substance. Irritant effects (clouding of the cornea, diffuse reddening and marked swelling of the conjunctiva) were found in 1 of the 3 treated rabbits, but were reversible after 3 days (Hoechst, 1983c). The test substance therefore caused slight irritation at most.

A further eye irritation study was carried out with 4-nitro-4'-aminodiphenylamine-2-sulfonic acid (58.1% pure, remainder water) in 3 adult female albino rabbits (strain HC:NZW, 2.9 to 3.6 kg) in accordance with OECD guideline no. 405. Each rabbit had 100 µl (ca. 50 mg) instilled into the conjunctival sac of one eye and the findings were assessed after 1, 24, 48 and 72 hours and after 7 days. The results were evaluated according to the Draize scheme. No irritant effects were found at any time (average irritation scores of 0 for the cornea, iris and conjunctiva). 4-Nitro-4'-aminodiphenylamine-2-sulfonic acid was therefore not irritating to the rabbit eye in this study (Bayer, 1986).

7.4   Sensitisation

The sensitising potential of 4-nitro-4'-aminodiphenylamine-2-sulfonic acid (60.6% pure; 3.4% dicondensation product, 0.8% paramine, 0.3% nitrochloric acid, 33.2% water) was investigated in a Magnusson and Kligman maximisation test in accordance with OECD guideline no. 406. The study was carried out in 20 female Dunkin-Hartley guinea pigs (average initial weight 385 g) with 20 further animals serving as controls. Based on the results of a preliminary study, intradermal induction was carried out with a 2.5% aqueous preparation, epidermal induction with a 25% aqueous preparation and the challenge treatment with 25 and 12.5% aqueous preparations. Formalin was used as a positive control. All 20 guinea pigs reacted to both the 12.5 and 25% preparations. 4-Nitro-4'-aminodiphenylamine-2-sulfonic acid was thus clearly shown to be a skin sensitiser in guinea pigs in this study (HRC, 1993).

7.5   Subchronic and chronic toxicity

No information available.

7.6   Genotoxicity

7.6.1 In vitro

4-Nitro-4'-aminodiphenylamine-2-sulfonic acid (58.5% pure, remainder water) was assessed in the Salmonella/microsome test in strains TA 100, TA 1535, TA 1537, TA 1538 and TA 98 and in Escherichia coli WP2uvrA with and without the addition of a metabolising system (S9-mix from Aroclor 1254-induced rat liver). It was found in a preliminary study that concentrations of 1000 or 2500 µg/plate were toxic to most of the bacterial strains. The concentrations used (4 to 5000 µg/plate) were chosen on the basis of this cytotoxicity test. Without metabolic activation the test substance was weakly mutagenic in TA 1538 only, while with metabolic activation a significant, concentration-dependent increase in revertants was seen in strains TA 1537, TA 1538 and TA 98 (Hoechst, 1984a).

In a further experiment, 4-nitro-4'-aminodiphenylamine-2-sulfonic acid (>99.7% pure) was studied in the Salmonella typhimurium strains TA 1538 and TA 98 with and without the addition of a metabolizing system (S9-mix from Aroclor 1254-induced rat liver). The concentrations tested ranged from 4 to 5000 µg/plate. Without metabolic activation, the substance showed a weak but concentration-dependent mutagenic effect in TA 1538; in the presence of a metabolising system, the substance caused significant concentration-dependent increases in revertants in both strains (Hoechst, 1984b).

From the original reports of both of the above studies, it can be concluded that the pure substance (>99.7%) is 1.4 to 1.9 times more effective than the technical grade substance (58.5%) at concentrations of 100 and 250 mg/plate, in comparable strains TA 1538 and TA 98. This is to some extent consistent with the "degree of dilution" of technical grade 4-nitro-4'-aminodiphenylamine-2-sulfonic acid. However, at higher concentrations, a reversal of this effect begins to emerge. It can also be inferred from the results that the contaminants (ca. 41.2%) in the technical grade 4-nitro-4'-aminodiphenylamine-2-sulfonic acid are not responsible for its mutagenic effect.

7.6.2 In vivo

No information available.

7.7   Carcinogenicity

No information available.

7.8 Reproductive toxicity

No information available.

7.9 Effects on the immune system

No information available.

7.10 Neurotoxicity

No information available.

7.11 Other effects

No information available.

8. Experience in humans

No information available.

9. Threshold limit values

No information available.

References

Bayer AG, Institut für Toxikologie
4-Nitro-4'-aminodiphenylamin-2-sulfonsäure   (Nitrophenyl-para-
minsäure), Untersuchungen zur akuten oralen Toxizität an männlichen und weiblichen Wistar-Ratten
Unpublished report (1981)

Bayer AG, Institut für Toxikologie
4-Nitro-4'-aminodiphenylamin-2-sulfonsäure, Untersuchungen zur
akuten oralen Toxizität an der Katze, Einfluß auf Methämoglo-
bingehalt und Zahl der HEINZ-Innenkörper im peripheren Blut
Unpublished report (1984)

Bayer AG, Institut für Toxikologie
4-Nitro-4'-aminodiphenylamin-2-sulfonsäure, Untersuchungen
zum Reiz-/Ätzpotential an Haut und Auge (Kaninchen)
Unpublished report no. 14308 (1986)

Bayer AG, Geschäftsbereich Organische Chemikalien
DIN-Sicherheitsdatenblatt   4-Nitro-4'-aminodiphenylamin-2-sul-
fonsäure (1989)

Ciba-Geigy Ltd., Basle, Switzerland
Report FAT 90142/A, acute dermal LD_{50} in the rat
Unpublished report project no. 831616 (1984)

Hoechst AG, Pharma Forschung Toxikologie
4-Nitro-4'-amino-diphenylamin-6-sulfosäure, Prüfung der akuten
oralen Toxizität an männlichen und weiblichen Wistar-Ratten
Unpublished report no. 83.0157 (1983a)

Hoechst AG, Pharma Forschung Toxikologie
4-Nitro-4'-amino-diphenylamin-6-sulfosäure, Prüfung auf akute
dermale Reizwirkung/Ätzwirkung am Kaninchen
Unpublished report no. 83.0142 (1983b)

Hoechst AG, Pharma Forschung Toxikologie
4-Nitro-4'-amino-diphenylamin-6-sulfosäure, Prüfung auf akute
Reizwirkung/Ätzwirkung am Auge beim Kaninchen
Unpublished report no. 83.0143 (1983c)

Hoechst AG, Pharma Research Toxicology
p-Nitrophenylparaminsäure TF, study of the mutagenic potential in
strains of Salmonella typhimurium (Ames test) and Escherichia coli
Unpublished report no. 84.0249 (1984a)

Hoechst AG, Pharma Research Toxicology
p-Nitrophenylparaminsäure TF, GER., study of the mutagenic
potential in strains of Salmonella typhimurium (Ames test), (TA
1538, TA 98)
Unpublished report no. 84.0668 (1984b)

Hoechst AG, Sicherheitsüberwachung
Written communication to BG Chemie of 16.06.1988.

Hoechst AG, Pharma Entwicklung Toxikologie
4-Nitro-4′-aminodiphenylamin-2-sulfonsäure, subakute orale Toxizität (28 Tage Fütterungsstudie) an SPF-Wistar-Ratten
Unpublished report no. 90.0987 (1990)
On behalf of BG Chemie

Hoechst AG, Sicherheitsüberwachung
Written communication to BG Chemie of 15.08.1991

HRC (Huntingdon Research Centre Ltd., Huntingdon, England)
No. 120 4-Nitro-4-aminodiphenylamine-2-sulfonic acid (CAS-no. 91–29–2), skin sensitisation in the guinea-pig
Unpublished report BGH 51/931584/SS (1993)
On behalf of BG Chemie

Triisobutyl phosphate

1. Summary and assessment

Based on the available animal data, triisobutyl phosphate is of low acute toxicity following oral administration (LD_{50} rat oral ca. 4200 mg/kg body weight and >5000 mg/kg body weight; rabbit dermal >5000 mg/kg body weight; 4-hour LC_{50} rat >5 mg/l air). No characteristic signs of toxicity occur in mice, rats, guinea pigs, rabbits or cats on repeated oral or inhalation exposure.

Triisobutyl phosphate has been reported to be both not irritating and moderately irritating to the skin, and is not irritating to the eye in the rabbit.

Triisobutyl phosphate is clearly sensitising to guinea pig skin in the maximisation test.

A 13-week feeding study in rats gave a no effect level of 1000 ppm (equivalent to 68 mg/kg body weight in males and 84 mg/kg body weight in females). No findings indicative of neurotoxicity were reported under the study conditions described.

Triisobutyl phosphate was not mutagenic in two Salmonella microsome tests which were carried out independently.

Triisobutyl phosphate did not display any embryotoxic or teratogenic activity in the rat.

In the domestic fowl, triisobutyl phosphate had no neurotoxic effect following two oral doses of 5000 mg/kg body weight administered 21 days apart with a total observation period of 42 days.

2. Name of substance

2.1   Usual name   Triisobutyl phosphate

2.2   IUPAC-name   Phosphoric acid triisobutyl ester

2.3   CAS-No.   126–71–6

2.4   EINECS-No   204–798–3

3.   Synonyms, common and trade names

Etingal
Phosphoric acid tris(2-methyl-
   propyl) ester
Triisobutylphosphat

4.   Structural and molecular formulae

4.1   Structural formula

$$
\begin{array}{l}
CH_3 \\
\quad\!\!\diagdown \\
\qquad CH-CH_2-O \\
\quad\!\!\diagup \\
CH_3 \\[4pt]
CH_3 \\
\quad\!\!\diagdown \\
\qquad CH-CH_2-O-P=O \\
\quad\!\!\diagup \\
CH_3 \\[4pt]
CH_3 \\
\quad\!\!\diagdown \\
\qquad CH-CH_2-O \\
\quad\!\!\diagup \\
CH_3
\end{array}
$$

4.2   Molecular formula $C_{12}H_{27}O_4P$

5.   Physical and chemical properties

5.1   Molecular mass, g/mol   266.32

5.2   Melting point, °C   <60   (BASF, 1991)

5.3   Boiling point, °C   272.5 (at 1013 hPa) (BASF, 1991)

5.4   Vapour pressure, hPa   2 (at 103 °C)
                                10 (at 133 °C)
                                50 (at 170 °C)   (BASF, 1991)

5.5   Density, g/cm³   0.9681 (at 20 °C)
                                             (Weast, 1979/80)

5.6 Solubility in water	264 mg/l (at 25 °C) (BASF, 1991)
5.7 Solubility in organic solvents	Soluble in ethanol, ether, benzene (Weast, 1979/80)
5.8 Solubility in fat	No information available
5.9 pH-value	-
5.10 Conversion factor	1 ml/m^3 (ppm)$\cong$10.87 mg/m^3 1 mg/m$^3\cong$0.09 ml/m^3 (ppm) (at 1013 hPa and 25 °C)

6. Uses

As a processing aid in paper and textile manufacture (BASF, 1991).

7. Experimental results

7.1 Toxicokinetics and metabolism

No information available.

7.2 Acute and subacute toxicity

In a preliminary study, an acute oral LD_{50} value of about 4.4 ml/kg, equivalent to about 4200 mg/kg body weight was determined in the rat (no further details; BASF, 1953).
Another source gives an acute oral LD_{50} of >5000 mg/kg body weight for the rat. The acute dermal LD_{50} in the rabbit was also >5000 mg/kg body weight (no further details; Monsanto, 1989).

Salivation and disturbances of balance occurred in the cat after a single oral dose of 1 ml/kg body weight (equivalent to about 960 mg/kg body weight; BASF, 1953).

The acute inhalation toxicity (LC_{50}) was determined in 5 male and 5 female rats (strain Hoe: WISKf/SPF71, average initial weights 205 and

188 g, respectively) in accordance with OECD guideline no. 403. The animals were exposed once (head-nose) for 4 hours. The analytically determined aerosol concentration was 5.14 mg/l air, with a mass-median aerodynamic particle diameter of 1.3 μm (geometric standard deviation 1.6). Signs of toxicity observed during the study included dyspnoea, trembling, uncoordinated movement, reduced spontaneous movement, ruffled fur, reduced righting and pinching reflexes, blood-coloured encrusted muzzles, sneezing, partly-closed eyes and red-stained salivary and nasal secretions. Body weight gain was reduced in the first week of observation. After 21 days, the rats were free of symptoms (except for one female), and body weight had returned to normal. No deaths occurred. At the end of the observation period, about half of the rats had discoloured lungs. The 4-hour LC_{50} for triisobutyl phosphate was thus >5.14 mg/l air (Hoechst, 1989a).

In earlier, preliminary acute inhalation toxicity studies, the results shown in Table 1 were obtained after exposure of laboratory animals to triisobutyl phosphate vapour or aerosols (BASF, 1953).

Table 1. Acute inhalation toxicity of triisobutyl phosphate (BASF, 1953)

Species	Nominal concentration	Duration of exposure	Mortality dead/exposed	Symptoms
Mouse	1.7 mg/l[1] (≙ 153 ppm)	6 hours	2/10	No pronounced symptoms
Rat	1.7 mg/l[1] (≙ 153 ppm)	6 hours	0/4	No signs of toxicity
Rabbit	1.7 mg/l[1] (≙ 153 ppm)	6 hours	0/1	No signs of toxicity
Cat	1.7 mg/l[1] (≙ 153 ppm)	6 hours	0/1	Salivation
Mouse	52 mg/l[2]	30 minutes	1/10	Pronounced dyspnoea
Rat	52 mg/l[2]	30 minutes	1/4	Pronounced dyspnoea
Rabbit	52 mg/l[2]	30 minutes	0/1	No signs of toxicity
Cat	52 mg/l[2]	30 minutes	0/1	Severe irritation of the mucous membranes associated with salivation, clouding of the cornea in the eye and lying on the side. These effects were reversible after exposure ceased.

1)   Vapour
2)   Aerosol

The administration of 0.5 ml triisobutyl phosphate/kg body weight/day by gavage (equivalent to 482 mg/kg body weight/day), on 22 days during a 30-day study, led to the death of the one rabbit used. At post-mortem, pea-sized focal haemorrhages were found in the lungs and the peritoneum was covered with a bloody-serous fluid. The blood supply to the gut was clearly increased, a few small haemorrhagic foci were found in the small intestine and the mucous membrane of the gut was inflamed (BASF, 1953).

In a further preliminary study in rabbits, the administration of 1 ml triisobutyl phosphate/kg body weight/day (964 mg/kg body weight/day) 3 times over 5 days by gavage caused the death of the animals. No clinical symptoms occurred after the first dose, while diarrhoea occurred after the second. Post-mortem examination revealed severe necrosis of the mucous membrane of the gut (BASF, 1953).

Findings on the oral toxicity of triisobutyl phosphate in the cat following repeated administration in a preliminary study are shown in Table 2.

Table 2. Toxicity of triisobutyl phosphate in the cat following repeated oral administration (1 animal/dose, preliminary study; BASF, 1953)

Dose	Days treated	Total treatment period (days)	Symptoms
0.05 ml/kg bw	23	39	Transitory diarrhoea
0.1 ml/kg bw	25	35	Loss of appetite, diarrhoea, reversible within 4-week observation period
0.1 ml/kg bw	38	52	Loss of appetite, diarrhoea
0.5 ml/kg bw	3	7	Salivation, unsteady gait, disturbances of balance, muscle tremors, ataxia, vomiting; symptoms intensified with repeated administration. Effects were reversible within 51-day observation period.
1.0 ml/kg bw	5	5	Salivation, disturbances of balance, diarrhoea, severe convulsions, death, doubling of leukocyte and granulocyte count, marked drop in lymphocyte count, protein in urine; no macroscopic effects at autopsy.

bw body weight

Table 3. Subacute inhalation toxicity of triisobutyl phosphate

Number/species	Effects (macroscopic)
10 Mice	Lethal in 1/10
4 Rabbits	No signs of toxicity
1 Rabbit	No signs of toxicity
1 Guinea pig	No signs of toxicity
1 Cat	No signs of toxicity

Table 3 shows the visible signs of toxicity evident in a preliminary study involving repeated inhalation of triisobutyl phosphate at a nominal concentration of between 0.3 and 1.0 mg/l (equivalent to 27 to 90 ppm) for 6 hours a day on 4 consecutive days (BASF, 1953).

7.3   Skin and mucous membrane effects

In an early study, triisobutyl phosphate was found to be "irritating" to the skin of rabbits (2 per treatment) both in concentrated form and as a 50% preparation in oil, while a 20% solution in oil gave rise only to a "suggestion of or a slight" irritant effect (20-hour occlusive application; BASF, 1953).

In other studies in rabbits, an average skin irritation score of 3.1 (maximum irritation score 8.0) was found for a single 4-hour exposure. Triisobutyl phosphate was therefore evaluated as moderately irritant (Monsanto, 1989).

A more recent skin irritation study was carried out in accordance with OECD guideline no. 404. Triisobutyl phosphate (0.5 ml, 90% pure, 10% n-propanol) was applied once to the clipped dorsal skin of 1 female and 2 male New Zealand white rabbits (weight 2.4 to 2.5 kg at the beginning of the study) for 4 hours under semi-occlusive cover. The effects were recorded one hour after exposure, after 24, 48 and 72 hours and after 7 days. Mild erythema (average irritation score 0.78), which was reversible after 7 days at most, was seen in all of the rabbits. Based on these results, the investigators evaluated triisobutyl phosphate as not irritating (RCC, 1990).

In the rabbit eye (2 animals in each case), the neat product caused marked irritant effects, while 50 and 20% solutions in oil caused only "a suggestion of irritation" (no further details; BASF, 1953).

In another eye irritation study in rabbits, triisobutyl phosphate caused mild to moderate temporary conjunctival irritation, which was reversible within one day in 4 of 6 rabbits and after 2 to 3 days in all rabbits. Triisobutyl phosphate was evaluated by the authors as not irritating (Monsanto, 1989).

7.4   Sensitisation

The sensitising potential of triisobutyl phosphate (purity unspecified) was tested in the maximisation test in guinea pigs (Pirbright-White strain, 10 test animals, 5 controls) in accordance with OECD guideline no. 406. Intradermal induction was carried out with the 0.05% product in paraffin DAB, while for dermal induction and dermal challenge, 4% triisobutyl phosphate in white vaseline DAB was used. The latter gave rise to a reaction in 9 of the 10 animals (Hoechst, 1989b). The test substance was therefore clearly sensitising in this study.

7.5   Subchronic and chronic toxicity

In a 13-week feeding study, groups of 10 male (170 to 218 g) and 10 female (132 to 174 g) Sprague-Dawley rats (CD) were given triisobutyl phosphate (99.7% pure) in the diet at concentrations of 0, 200, 1000 and 5000 ppm ($\pm$10% in each case). This was equivalent to a daily substance intake of 0, 13.9, 68.4 and 364.1 mg/kg body weight in the males and 0, 16.8, 84.3 and 403.9 mg/kg body weight in the females. A further 10 male and 10 female rats were used in each of the control and top dose groups. These were changed to a normal diet after 13 weeks before being killed after an 8-week observation period. No deaths or treatment-related signs of toxicity occurred. Nothing abnormal was detected on ophthalmoscopic examination. There were no differences in body weight gain during the entire duration of the study. Males at the top dose level showed a significant reduction in neutrophilic leukocytes and significantly higher average corpuscular haemoglobin while males in the intermediate and high dose groups showed a significant increase in average corpuscular haemoglobin concentration. A significant increase in the cholesterol level seen in the males at the top dose was considered to be a possible treatment-related effect. No observations were made regarding whether these effects were reversible. Post-mortems and histological examination of ca. 40 organs revealed no treatment-related effects. The no effect level was given as 1000 ppm, equivalent to 68.4 mg/kg body weight/day in the males and 84.3 mg/kg body weight/day in the females (Monsanto, 1990). No indications of neurotoxicity were described under the study conditions employed.

In the rat, the oral administration of a saturated aqueous solution of triisobutyl phosphate (about 440 and 430 mg/kg body weight in males and females, respectively) for 51 days, in a drinking water study, led to a slight reduction in body weight gain compared with the controls. No gross effects were evident at autopsy (BASF, 1953).

7.6　Genotoxicity

7.6.1 In vitro

Triisobutyl phosphate (ca. 98% pure) was studied in the Salmonella/microsome test in strains TA 1535, TA 100, TA 1537 and TA 98 with and without metabolic activation (S9-mix from Aroclor 1254-induced rat liver). The studies were carried out with the standard plate test (SPT) and the preincubation test (PIT) at concentrations of 20 to 5000 μg/plate (SPT) and 15 to 5000 μg/plate (PIT). In the SPT, concentrations of $\geq$ 2500 μg/plate and $\geq$ 100 μg/plate were cytotoxic in strains TA 100 and TA 1535, respectively, while in the PIT, concentrations of $\geq$ 500 μg/plate were toxic to all strains. Under these study conditions, triisobutyl phosphate did not show any mutagenic activity either with or without metabolic activation (BASF, 1990).

A further Salmonella/microsome test was carried out in strains TA 1535, TA 1537, TA 1538, TA 98 and TA 100 with and without metabolic activation (S9-mix from Aroclor 1254-induced rat liver). A plate incorporation test and a preincubation test were carried out, both at triisobutyl phosphate concentrations of 10, 33.3, 100, 333.3, 1000 and 5000 μg/plate. Cytotoxicity occurred in nearly all strains at the higher concentrations both with and without metabolic activation. Triisobutyl phosphate had no mutagenic activity in these studies either with or without metabolic activation (CCR, 1992).

7.6.2 In vivo

No information available.

7.7   Carcinogenicity

No information available.

7.8   Reproductive toxicity

In a reproductive toxicity study, groups of 25 female Charles River rats (178 to 223 g) were given 0, 100, 300 or 1000 mg ca. 98% pure triisobutyl phosphate/kg body weight as an aqueous formulation with 0.5% carboxymethylcellulose sodium from day 6 to day 15 of pregnancy. Following surgical delivery on day 20, the dams and the foetuses were examined, about half of the foetuses being used for the detection of soft-tissue malformations and half for skeletal malformations. The 1000 mg/kg dose caused marginal maternal toxicity with incipient reductions in body weight gain, possibly increased water intake and salivation following gavaging. The latter was also observed in rats in the 300 and 100 mg/kg dose groups. There was a marginal increase in early resorptions in the rats in the top dose group (1000 mg/kg), although only a few litters were affected and the effect was not significant. Triisobutyl phosphate had no effect on embryonic or foetal development and did not cause significant visceral or skeletal anomalies or skeletal variations. A slight increase in bilateral curvature of the forelegs was limited to two litters in the top dose group and was not considered to be treatment-related (HRC, 1991).

7.9   Effects on the immune system

No information available.

7.10 Neurotoxicity

A study of the possible neurotoxic effect of triisobutyl phosphate (99.7% pure) was carried out based on OECD guideline no. 418, in groups of 5 adult domestic hens (aged ca. 12 to 14 months). Two groups were given single doses of 500 and 1000 mg/kg body weight, respectively, (dissolved in olive oil) by gavage and were then observed for 42 days. Further groups were given single doses of 2000 and 5000 mg/kg body weight, respectively, and were then given the same dose again 21 days later with a total observation period of 42 days. As a positive control, 3 groups of 5 hens were given a single dose of 500 mg tri-o-cresyl phosphate/kg body weight by gavage (also in olive oil) and were also observed for 42 - days. In the positive control animals, treatment-related deaths (5 of 15), serious or severe neurotoxic symptoms (7 of 15) and weight loss occurred, while in the triisobutyl phosphate-treated hens there were no effects that could be attributed to treatment, even after the administration of two doses of 5000 mg/kg body weight (BASF, 1987).

Further studies on the possible neurotoxic effects of triisobutyl phosphate involved the determination of specific enzymes in hens. A single oral dose of 5000 mg/kg body weight of the neat product was given to 10 adult hens (ca. 1.6 kg, about 17 months old). Further groups of 5 hens remained untreated or were given a single dose of 750 mg tri-o-cresyl phosphate/kg body weight as a positive control. Following treatment, the activities of acetylcholinesterase and neurotoxic esterase in the brain and of butyrylcholinesterase in the serum were measured. Neurotoxic esterase activity serves as a marker of organophosphate-induced delayed neurotoxicity. Triisobutyl phosphate did not reduce the activities of acetylcholinesterase or neurotoxic esterase in the brain (93 and 100% of untreated control values, respectively), but did decrease serum butyrylcholinesterase activity (2% of untreated control). After administration of tri-o-cresyl phosphate, the activities were 79% (acetylcholinesterase), 14% (neurotoxic esterase) and 11% (butyrylcholinesterase). These results led the authors to conclude that triisobutyl phosphate did not cause delayed neurotoxicity in hens. Details of neurotoxic symptoms were not provided (Laboratory of Neurotoxicology, 1990).

7.11 Other effects

No information available.

8.   Experience in humans _____________________________________

No information available.

9.   Threshold limit values _____________________________________

No information available.

References _____________________________________

BASF AG, Abteilung Toxikologie
Unpublished report (1953)

BASF AG, Abteilung Toxikologie
Orientierende Prüfung auf akute verzögerte neurotoxische Wirkung von Etingal (Tri-iso-butylphosphat) am Haushuhn (Gallus gallus gallus L.)
Unpublished report, project no. 95WO175/84 (1987)

BASF AG, Abteilung Toxikologie
Report on the study of Etingal (ZST Test Substance No.: 90/404) in the Ames test (standard plate test and preincubation test with Salmonella typhimurium).
Unpublished report, project no. 40M0404/904191 (1990)

BASF AG
AIDA-Grunddatensatz Phosphoric acid, tris(2-methylpropyl) ester (1991)

CCR (Cytotest Cell Research GmbH & Co. KG, Roßdorf)
Salmonella typhimurium reverse mutation assay - plate incorporation test, preincubation test - with triisobutylphosphate (no. 112)
Unpublished report, CCR project 139500 (1992)
On behalf of BG Chemie

Hoechst AG, Pharma Forschung Toxikologie und Pathologie
Triisobutylphosphat, Prüfung der akuten Aerosolinhalationstoxizität
an männlichen und weiblichen SPF-Wistar Ratten, 4-Stunden- LC_{50}
Unpublished report no. 89.1632 (1989 a)
On behalf of BG Chemie

Hoechst AG, Pharma Forschung Toxikologie und Pathologie
Triisobutylphosphat, Prüfung auf sensibilisierende Eigenschaften an
Pirbright-White-Meerschweinchen im Maximierungstest
Unpublished report no 89.0587 (1989 b)
On behalf of BG Chemie

HRC (Huntingdon Research Centre Limited, Huntingdon, UK)
A study of the effect of tri-isobutylphosphate (no. 112, CAS-no.:
126–71–6) on pregnancy of the rat
Unpublished report no. BGH 24/18/901761 (1991)
On behalf of BG Chemie

Laboratory of Neurotoxicology, Department of Pharmacology,
Duke University Medical Center, Durham, North Carolina
Study on the delayed neurotoxicity of tri-iso-butyl phosphate
Report, study no. DU-89–470 (1990)
On behalf of the Monsanto Company
NTIS/OTS 0528544

Monsanto, Department of Medicine and Health Sciences
Acute toxicity and irritation studies with tri-isobutylphosphate
Unpublished report no. BD-88–400 (1989)

Monsanto Agricultural Company, Environmental Health Laboratory
90-day study of triisobutyl phosphate (TIBP) administered in feed to
albino rats
Report, Project no. ML-89–460 (1990)
NTIS/OTS 0528780

RCC (Research & Consulting Company AG)
Primary skin irritation study with Tri-iso-butylphosphat in rabbits (4-
hour semi-occlusive application)
Unpublished report, RCC project 226844 (1990)
On behalf of BG Chemie

Weast, R.C.
CRC Handbook of chemistry and physics
60th ed., C-435
CRC Press, Boca Raton, Florida (1979/80)

Antimony-V-oxide

1. Summary and assessment

Following intraperitoneal injection in the rat, 88% of pentavalent antimony is excreted in the urine and 1% in the faeces.

The acute oral toxicity (LD_{50}) of antimony-V-oxide in dogs was >400 mg/kg body weight.

No signs of toxicity occurred in preliminary studies in a rat, involving repeated administration in the diet (an increase in the daily dose over 9 days from 100 to 2000 mg/rat/day); similarly, the administration of increasing doses of from 0.1 to 4 mg/rat/day over 107 days was without overt effects.

Antimony-V-oxide was not genotoxic in the Salmonella/microsome test in strains TA 98 or TA 100 either with or without metabolic activation, in the spot test in *Bacillus subtilis* or in a sister chromatid exchange study. The last two results however, are of limited value due to the low solubility of antimony-V-oxide.

In humans, pneumoconiosis has been described after chronic exposure, although the cases involved mixed exposure, with antimony-V-oxide present at only 2 to 8%.

Biomonitoring of workers exposed to pentavalent antimony can be carried out by determination of the antimony content of the urine.

2. Name of substance

2.1	Usual name	Antimony-V-oxide
2.2	IUPAC-name	Antimony-V-oxide
2.3	CAS-No.	1314–60–9
2.4	EINECS-No.	215–237–7

3.   Synonyms, common and trade names

Antimonic "acid"
Antimonic oxide
Antimon-V-oxid
Antimony pentaoxide
Antimony pentoxide
Diantimony pentaoxide
Diantimony pentoxide
Stibic anhydride

4.   Structural and molecular formulae

4.1   Structural formula

Sb_2O_5 (polymer), octahedral, coordinate bonds to oxygen throughout
(Holleman-Wiberg, 1985)

4.2   Molecular formula

Sb_2O_5

5.   Physical and chemical properties

5.1   Molecular mass, g/mol

323.50

5.2   Melting point, °C

No information available

5.3   Boiling point, °C

–

5.4   Vapour pressure, hPa

No information available

5.5   Density, g/cm^3

3.78   (Windholz et al., 1983)

5.6   Solubility in water

Very low solubility
(Holleman-Wiberg, 1985)

5.7   Solubility in organic solvents

No information available

5.8   Solubility in fat

No information available

5.9   pH value   –

5.10 Conversion factor   1 ml/m^3(ppm) $\triangleq$ 13.20 mg/m^3
                                            1 mg/m^3 $\triangleq$ 0.076 ml/m^3(ppm)
                                            (at 1013 hPa and 25 °C)

6.   Uses

As a fire retardant (Windholz et al., 1983).

7.   Experimental results

7.1   Toxicokinetics and metabolism

Male Sprague-Dawley rats were each given a single intraperitoneal injection of 3 μg pentavalent [124]antimony. Of the administered radioactivity, 94 and 96% was bound to the blood plasma in dialysable form after 45 minutes and 8 hours, respectively. 88% of the radioactivity was excreted in the urine and only 1% in the faeces within 24 hours. Following a single intravenous injection of 0.1 μg pentavalent antimony/rat, ca. 30% was excreted in the urine and about 1.3% in the bile within 140 minutes (Edel et al., 1993).

7.2   Acute and subacute toxicity

The following LD$_{50}$ values have been determined after single intra-peritoneal or oral doses of antimony-V-oxide.

Mouse	i. p.	978 mg/kg body weight (TPI, 1982)
Rat	i. p.	4000 mg/kg body weight (Bradley and Fredrick, 1941; Venugopal and Luckey, 1978)
Dog	p. o.	>400 mg/kg body weight (Flury, 1927)

Information on signs of toxicity observed was not provided.
Groups of 12 female albino rats were each given a single dose of 50 mg antimony oxide dust as an ethanolic-aqueous suspension

(0.5 ml) either by intraperitoneal injection or by intratracheal instillation. The material studied consisted of antimony dust deposits from a foundry. The dust contained 38.73 to 88.86% Sb_2O_3, 2.11 to 7.82% antimony-V-oxide, 0.90 to 3.81% Fe_2O_3, 0.82 to 4.72% SiO_2 and 0.21 to 6.48% As_2O_3. On histological examination of the peritoneum and the lungs 2 months after dosing, pneumoconiosis was detectable ; the collagen was unaffected (number of animals affected unspecified; Potkonjak and Vishnjich, 1983).

In a preliminary study, a rat was given ca. 100 mg antimony-V-oxide/day in the diet, increasing to 2000 mg/day within 9 days. The behaviour of the rat was unaffected, and the faeces were of a normal consistency but orange-yellow in colour (Flury, 1927).

A cat was given 20 mg/day in the diet, increasing to 600 mg/day over 29 days. There followed a 5-day observation period. Vomiting and diarrhoea occurred on day 19 and was also observed on the following days. On days 27 and 28, severe vomiting again occurred. Feed intake was markedly reduced and body weight gain inhibited. The symptoms cleared up completely during the 5-day observation period (Flury, 1927).

7.3   Skin and mucous membrane effects

No information available.

7.4   Sensitisation

No information available.

7.5   Subchronic and chronic toxicity

In a preliminary study, a rat (ca. 40 g at the beginning of the study) was given 0.1 mg antimony-V-oxide/day in the diet for 21 days followed by 0.2 mg/day for 56 days, with a subsequent 25-day observation period. There were no adverse effects on behaviour (Flury, 1927).

A further rat (ca. 25 g at the beginning of the study) was given 0.1 mg antimony-V-oxide/day in the diet for 21 days followed by 0.2 mg/day for 56 days. The dose was then increased to 0.4 mg/day for 7 days, 0.6 mg/day for 7 days, 0.8 mg/day for 7 days, 2.0 mg/day for 4 days and 4.0 mg/day for 5 days. There were no adverse effects on behaviour (Flury, 1927).

7.6   Genotoxicity

7.6.1 In vitro

Antimony-V-oxide (99.9% pure) was not mutagenic in a Salmonella/microsome preincubation test in strains TA 98 and TA 100 at concentrations of 50, 100 and 200 μg/plate with and without metabolic activation (S9-mix from rat liver, no details of induction). Other strains were not tested (Kuroda et al., 1991; Endo et al., 1991).

In a spot test in *Bacillus subtilis* H17/M45, antimony-V-oxide (99.9% pure) was not genotoxic at a dose of 60 mg/plate. However, the authors did not exclude the possibility of a DNA-damaging effect, as higher doses could not be tested due to the low solubility in water (Kuroda et al., 1991).

Antimony-V-oxide (99.9% pure) was not genotoxic in a sister chromatid exchange test in Chinese hamster V79 cells at concentrations of 10, 20 and 40 μg/ml. However, the cytotoxic range could not be determined due to the low solubility in water (Kuroda et al., 1991; Endo et al., 1991).

Potassium antimonate has also been tested for genotoxic activity in the standard plate incorporation test and in the fluctuation test in the Salmonella strains TA 98, TA 100, TA 1535, TA 1537 and TA 1538 and in Escherichia coli WP2uvrA. The substance was negative in the incorporation test but gave equivocal results in the fluctuation test. These general statements were made however, without the presentation of individual results (Arlauskas et al., 1985).

7.6.2 In vivo

No information available.

7.7   Carcinogenicity

No information available.

7.8   Reproductive toxicity

No information available.

7.9   Effects on the immune system

No information available.

7.10   Neurotoxicity

No information available.

7.11   Other effects

No information available.

8.   Experience in humans

Pneumoconiosis developed in 51 foundry workers (31 to 54 years old) who had been exposed to antimony oxide dust for between 9 to 31 years (average 17.91 years). The dust contained 38.73 to 88.86% Sb_2O_3, 2.11 to 7.82% antimony-V-oxide, 0.90 to 3.81% Fe_2O_3, 0.82 to 4.72% SiO_2 and 0.21 to 6.48% As_2O_3. Concentrations ranged from 17 to 86 mg dust/m^3 air depending on the production process. The number of dust particles was between 1380 and 5988/m^3 air. Clinical symptoms included chronic cough (60.8 % of workers studied), conjunctivitis (27.5%), inflammation of the upper airways (35.8%) and "antimony dermatosis" (62.8%), particularly during summertime, with vesicular and occasionally pustulous outbreaks leaving residual hyperpigmentation. Shadows on the lungs were found on X-ray examination. Conditions diagnosed in the cohort included chronic bronchitis (37.3%), chronic emphysema (34.5%), latent tuberculosis (18.2%) and pleural adhesions (27.3%). The cardiovascular, biochemical and haematological parameters were within normal limits, and no impairment of the central or peripheral nervous system was observed (Potkonjak and Pavlovich, 1983).

A group of 22 workers involved in the production of antimony pentoxide and sodium antimonate was equipped with personal air samplers for one or two shifts in order to determine the concentration of antimony-V. At the beginning and the end of the working day, the urine was analyzed for its antimony and creatinine content and the samples divided according to the production processes used ("wet"/"dry"). In the workers involved in the wet process, the antimony concentration in the urine rose from 8.2 μg/g creatinine before the shift to 12.3 μg/g creatinine after the end of the shift, at a concentration of 86 μg/m^3 air. In the workers involved in the dry process, the corresponding values were 58.4 and 110 μg/g creatinine, respectively, at a concentration in air of 927 μg/m^3. Regression analysis showed that an air concentration of 500 μg/m^3 during the shift, the current threshold value, would lead to a concentration of 35 mg antimony/g creatinine. The determination of pentavalent antimony in the urine was therefore recommended for monitoring exposure (Bailly et al., 1991).

9.   Threshold limit values

The MAK value for antimony (Sb) in the Federal Republic of Germany is 0.5 mg total Sb dust/m^3 air (DFG, 1993; TRGS, 1994).
The TLV-TWA value in the USA is also 0.5 mg Sb/m^3 air (ACGIH, 1993–1994).

References

ACGIH (American Conference of Governmental Industrial Hygienists)
Threshold limit values for chemical substances and physical agents and biological exposure indices (1993–1994)

Arlauskas, A., Baker, R.S.U., Bonin, A.M., Tandon, R.K., Crisp, P.T., Ellis, J.
Mutagenicity of metal ions in bacteria
Environ. Res., 36, 379–388 (1985)

Bailly, R., Lauwerys, R., Buchet, J.P., Mahieu, P., Konings, J.
Experimental and human studies on antimony metabolism: their
relevance for the biological monitoring of workers exposed to inorganic antimony
Br. J. Med., 48, 93–97 (1991)

Bradley, W.R., Fredrick, W.G.
The toxicity of antimony – animal studies
Ind. Med. Chicago, 2, 15–22 (1941)

DFG (Deutsche Forschungsgemeinschaft)
Maximale Arbeitsplatzkonzentrationen und Biologische Arbeits-
stofftoleranzwerte 1993
VCH Verlagsgesellschaft, Weinheim (1993)

Edel, J., Marafante, E., Sabbioni, E., Manzo, L.
Metabolic behaviour of inorganic forms of antimony in the rat
International Conference on Heavy Metals in the Environment,
574–577 (1993)

Endo, G., Kuroda, K., Okamoto, A., Yoo, Y.S., Horiguchi, S.
Genotoxicity of Be, Ga, Sb and As compounds
Mutat. Res., 252 (1), 84–85 (1991)

Flury, F.
Zur Toxikologie des Antimons
Arch. Exp. Pathol. Pharmakol., 126, 87–103 (1927)

Herbst, K.A., Rose, G., Hanusch, K., Schumann, H., Wolf, H.U.
Antimony and antimony compounds
In: Ullmann's encyclopedia of industrial chemistry
5th ed., vol. A3, pp. 55–76
VCH Verlagsgesellschaft, Weinheim (1985)

Holleman-Wiberg
Lehrbuch der Anorganischen Chemie
91st – 100th ed., p. 688–689
Walter de Gruyter, Berlin, New York (1985)

Kuroda, K., Endo, G., Okamoto, A., Yoo, Y.S., Horiguchi, S.
Genotoxicity of beryllium, gallium and antimony in short-term assays
Mutat. Res., 264, 163–170 (1991)

Potkonjak, V., Pavlovich, M.
Antimoniosis: a particular form of pneumoconiosis. I. Etiology, clinical and X-ray findings
Int. Arch. Occup. Environ. Health, 51, 199–207 (1983)

Potkonjak, V., Vishnjich, V.
Antimoniosis: a particular form of pneumoconiosis.
II. Experimental investigation
Int. Arch. Occup. Environ. Health, 51, 299–303 (1983)

RTECS (Registry of Toxic Effects of Chemical Substances)
Antimony pentoxide, RTECS-no. CC6300000
produced by NIOSH (National Institute for Occupational Safety and Health) (1990)

TPI (Toxic Parameter Index)
Toxic chemicals under single exposure, p. 23 (1982)
Cited in: RTECS (1990)

TRGS (Technische Regeln für Gefahrstoffe) 900
Grenzwerte in der Luft am Arbeitsplatz
Bundesarbeitsblatt, 6, p. 37 (1994)

Venugopal, B., Luckey, T.D.
Metal toxicity in mammals
Vol. 2 Chemical toxicity of metals and metalloids, pp. 213–215
Plenum Press, New York and London (1978)

Windholz, M., Budavari, S., Blumetti, R.F., Otterbein, E.S. (eds.)
The Merck index
10th ed., p. 103
Merck & Co., Inc., Rahway, USA (1983)

Dimethylaminopropionitrile

1. Summary and assessment

After oral administration of dimethylaminopropionitrile, rats excrete the substance in the urine unchanged, together with β-aminopropionitrile and cyanoacetic acid. In vitro, formaldehyde, cyanoacetic acid and cyanide ions can be detected after incubation of dimethylaminopropionitrile with liver, kidney and bladder homogenates, with cyanoacetaldehyde suggested as the reactive metabolite. Metabolism appears to proceed primarily through the cytochrome P-450-dependent mixed-function oxidases.

Dimethylaminopropionitrile is of low to moderate toxicity after oral administration to mice (LD_{50} 1500 mg/kg body weight) and rats (LD_{50} 1293 to 2240 mg/kg body weight) and is moderately toxic to rabbits on dermal exposure (LD_{50} 1212 mg/kg body weight).

The predominant signs of toxicity in rats and mice after acute exposure are central excitation and disorders of micturition. Oral administration of 350 mg dimethylaminopropionitrile/kg body weight to rats and mice twice within 24 hours leads to histopathological changes in the kidneys and bladder within a few hours. The same effects have been described in rats on oral administration for 2 weeks. In contrast to the metabolite, β-aminopropionitrile, dimethylaminopropionitrile (0.35% in the diet) causes no neuro- or osteolathyrism.

Dimethylaminopropionitrile is not irritating to the skin, but is irritating to the eyes of rabbits.

In the Salmonella/microsome test, dimethylaminopropionitrile is not mutagenic, either with or without metabolic activation.

Dimethylaminopropionitrile had no embryotoxic activity in rats in a study that was inadequately reported and not carried out in accordance with current guidelines.

Experience in man has not been gained with the pure substance, but with the catalyst ESN-NIAX, which consists of up to ca. 95% dimethylaminopropionitrile. Information is available on the effects of both inhalation and dermal exposure. According to American observations, the major symptom of intoxication is a neurological dysfunction of the urinary bladder. In addition, disturbances in sexual behaviour, paraesthesia and weakened reflexes in the extremities, skin lesions and irritation of the respiratory tract have been reported. Disturbances of conduction (involving the sensory pathway of the peripheral nerves) have also been found. The symptoms are largely reversible after exposure ceases, although isolated cases of difficulty in urinating and sexual dysfunction have been reported 2 years after the end of exposure.

2.   Name of substance

2.1 Usual name	Dimethylaminopropionitrile
2.2 IUPAC-name	3-(Dimethylamino)-propionitrile
2.3 CAS-No.	1738–25–6
2.4 EINECS-No.	217–090–4

3.   Synonyms, common and trade names

β-N-Diaminopropionitrile
β-Dimethylaminopropionitrile
Dimethylaminopropionitril
N,N-Dimethylamino-
  3-propionitrile
DMAPN
3-(Dimethylamino)-
  propanenitrile
ESN-NIAX
Nitrile Z
NIAX-ESN
NIAX catalyst ESN
Propionitrile, 3-(dimethylamino)-

4.   Structural and molecular formulae

4.1   Structural formula

$$\begin{array}{c} H_3C \\ \searrow \\ N\text{-}CH_2\text{-}CH_2\text{-}CN \\ \nearrow \\ H_3C \end{array}$$

4.2   Molecular formula $C_5H_{10}N_2$

5.   Physical and chemical properties

5.1	Molecular mass, g/mol	98.15	
5.2	Melting point, °C	– 44.2	(Hartung, 1982)
5.3	Boiling point, °C	172	(Hartung, 1982)
5.4	Vapour pressure, hPa	2.4 (at 30 °C)	(BASF, 1992)
		13.3 (at 57 °C)	(Hartung, 1982)
5.5	Density, g/cm^3	0.8617 (at 30 °C)	
			(Hartung, 1982)
5.6	Solubility in water	Miscible	(Hartung, 1982)
5.7	Solubility in organic solvents	Miscible with alcohol	(Hartung, 1982)
5.8	Solubility in fat	No information available	
5.9	pH value	No information available	
5.10	Conversion factor	1 ml/m^3 (ppm) $\cong$ 4.01 mg/m^3	
		1 mg/m^3 $\cong$ 0.25ml/m^3 (ppm)	
		(at 1013 hPa and 25 °C)	

6.   Uses

Catalyst in the manufacture of polyurethane foam, intermediate in the manufacture of N,N-dimethyl-1,3-diaminopropane and other organic compounds (NIOSH, 1978).

7.   Experimental results

7.1   Toxicokinetics and metabolism

In vivo studies have shown that following the administration of a single oral dose of 4.4 mmol dimethylaminopropionitrile/kg body weight to rats, β-aminopropionitrile and cyanoacetic acid were excreted in the urine, but cyanide was not. There was a linear correlation between the toxicity of dimethylaminopropionitrile (dysfunction of the urinary bladder) and its metabolism to cyanoacetic acid. Equimolar doses of cyanoacetic acid had similar toxic effects to dimethylaminopropionitrile in 50% of the treated animals. In vitro studies have shown that homogenates of liver, kidney or bladder metabolise dimethylaminopropionitrile to formaldehyde, cyanoacetic acid and cyanide ions, with the formation of cyanoacetic acid probably occurring via the highly reactive cyanoacetaldehyde. From this the authors concluded that cyanoacetaldehyde is a possible toxic metabolite of dimethylaminopropionitrile (Ahmed and Farooqui, 1984).

Liver cell microsomes and rat liver tissue were incubated with 10 μmol dimethylaminopropionitrile/3.0 ml incubation medium for one hour at 37 °C, and the cyanide content was then measured. This was 15.4 (±2.9) nmol/10 mg protein in the microsomal suspension and 42.7 (±2.3) nmol/500 mg wet material for the liver tissue. There were also indications of formaldehyde formation (0.32± 0.06 μmol) in the in vitro studies in microsomes and liver tissue. Cyanide could also be detected as thiocyanate in the urine of rats in vivo, following intraperitoneal injection. Male Sprague-Dawley rats (180 to 220 g) were given a single intraperitoneal injection of 174 mg dimethylaminopropionitrile/kg body weight, and their collected urine was analyzed 48 hours after dosing. The thiocyanate content was 8.2±0.9 mg/kg, equivalent to 8.0±0.2% of the administered dose after subtraction of endogenously formed thiocyanate (Froines et al., 1985).

5 male Sprague-Dawley rats (175 to 225 g) were each treated with 525 mg (unlabelled, 98% pure) dimethylaminopropionitrile/kg body weight by gavage twice within 24 hours. Within 5 days, ca. 44% of the administered dose had been excreted in the urine as unchanged dimethylaminopropionitrile, about half of this within

the first 12 hours. β-Aminopropionitrile and cyanoacetic acid were identified as metabolites. In vitro studies revealed that further metabolism of dimethylaminopropionitrile to cyanide, formaldehyde and cyanoacetic acid mainly occurs in the microsomal frac-

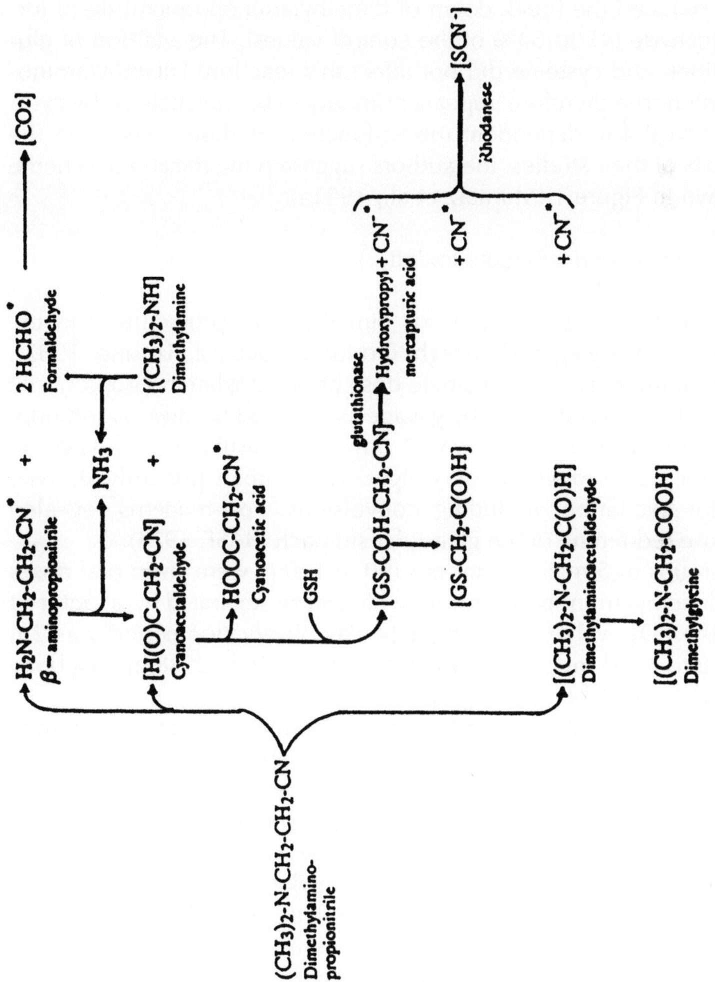

Fig. 1. Suggested scheme for the metabolism of dimethylaminopropionitrile (Mumtaz et al., 1991a)

tion of the liver, the kidneys and the bladder. These reactions are dependent on NADPH and oxygen for maximum activity. Liver microsomes induced with pentobarbital in vivo increased dimethylaminopropionitrile metabolism (220% of control value) and in vivo pretreatment with cobalt chloride reduced it (73% of control value). In vitro pretreatment with SKF 525-A (a cytochrome P-450 inhibitor) reduced the break down of dimethylaminopropionitrile to formaldehyde (47 to 64% of the control values). The addition of glutathione and cysteine did not affect this reaction. Dimethylaminopropionitrile therefore appears primarily to be metabolised by cytochrome P-450-dependent mixed-function oxidases. Based on the results of their studies, the authors suggested the metabolic scheme shown in Figure 1 (Mumtaz et al., 1991a).

7.2   Acute and subacute toxicity

The acute oral LD_{50} value for dimethylaminopropionitrile in the mouse is 1.5 g/kg body weight (no further details; Hartung, 1982). In a preliminary study, a single dose of dimethylaminopropionitrile was administered to rats by gavage as a 6 to 33% aqueous solution. The observation period was 7 days. The resulting LD_{50} was ca. 1.5 ml (1290 mg)/kg body weight. The animals primarily showed motor excitation, including convulsions. Post-mortems revealed diffuse reddening of the glandular stomach (BASF, 1974).

Groups of 5 male Wistar rats (90 to 120 g) were given oral doses of dimethylaminopropionitrile which were separated by a factor of 2. The LD_{50} value following a 14-day observation period was 2.6 (1.98 to 3.41) ml/kg body weight (2240 (1706 to 2938) mg/kg body weight; no further details; Smyth et al., 1962).

The same authors investigated the dermal toxicity of dimethylaminopropionitrile. Increasing doses of the test substance were applied to the clipped skin of groups of 4 male New Zealand white rabbits (initial weights 2.5 to 3.5 kg) and covered with film for 24 hours. The LD_{50} was calculated to be 1.41 (0.88 to 2.29) ml/kg body weight (1212 (757 to 1969) mg/kg body weight) after an observation period of 14 days (Smyth et al., 1962).

It has also been reported that the acute LD_{50} for dimethylaminopropionitrile is 385 mg/kg body weight following intraperitoneal injection in the mouse and 445 µl (383 mg)/kg body weight following oral administration in the rabbit (no further details; UCC, 1978; Vitorskii, 1961).

A further intraperitoneal LD_{50} value of ca. 0.5 ml (430 mg)/kg body weight (with an aqueous solution) was given for the mouse. The observation period was 7 days. Signs of toxicity observed were motor excitation and convulsions. There were no conspicuous findings at autopsy (BASF, 1974).

The acute toxicity of the catalyst ESN-NIAX, which contains ca. 95% dimethylaminopropionitrile (other constituents ca. 5% bis(2-dimethylamino)ethyl ether, 0 to 0.05 % acrylonitrile, 0 to 0.2% dimethylamine), was tested in male Holtzmann rats (305 to 340 g; 3/group). The rats were given the catalyst as a 20% solution in distilled water at doses of 0.2 and 2.0 ml/kg body weight by intraperitoneal injection. Further rats (3/group) were given either 0.5 or 1.0 ml of a 20% aqueous solution by gavage (equivalent to ca. 0.31 and 0.62 ml ESN-NIAX/kg body weight). This dose was given twice within 24 hours and the rats were killed ca. 8 hours later. The bladder, liver, lungs, spleen, kidneys, adrenal glands, stomach and duodenum were examined histologically. Intraperitoneal injection of 2.0 ml/kg led to the death of the rats with tonic spasms within one minute. Injection of 0.2 ml/kg caused a reduction in motor activity, accelerated breathing and lying on the side. In the surviving rats (number not reported), these symptoms were reversible within 10 to 15 minutes. No comparable signs of toxicity were seen on oral administration of the catalyst. Enlarged bladders, in which histological examination revealed transmural oedema, acute ulcers, inflammation (particularly in the submucosa) and desquamation of necrotic cells, were evident at post-mortem (8 hours after the last dose). The inflammation had in some cases spread to the prostate. Haemorrhagic erosions of the gastric mucosa were diagnosed (Jaeger et al., 1980).

In an inhalation hazard test, exposure to an atmosphere saturated with dimethylaminopropionitrile at 20 °C for 3 hours was tolerated by 12 rats, while 1 of 12 rats exposed for 8 hours died. Signs of toxicity included disturbances of balance, pronounced secretion from the eyes and nose and severe trembling. Autopsy revealed acute dilation of the heart and hyperaemia of the lungs (BASF, 1974).

Groups of 5 male Swiss-Webster mice (27.8 to 36 g) were given intraperitoneal injections of 1.0, 0.5 or 0.25 ml dimethylamino-propionitrile or ESN-NIAX/kg body weight 5 days/week until death occurred or for a maximum of 2 weeks. A control group of 5 mice was treated with physiological saline. Animals that survived the treatment were observed for a week. Histological examinations were carried out in 2 control and 2 surviving mice. Mice at the 1.0 and 0.5 ml/kg dose levels died within a few minutes of injection. Both test substances caused lacrimation, tremors and tonic convulsions. In the low-dose group, convulsions and tremors occurred within the first 3 days of dimethylaminopropionitrile administration. The same symptoms were observed after administration of the lowest dose of the catalyst, together with weight loss and generalised malaise. The condition of the animals returned to normal during the observation period. Microscopic examination of the tissues did not reveal any abnormalities (Gad et al., 1979).

A study of the same design as the mouse study described above was carried out in groups of 5 male Sprague-Dawley rats (180 to 200 g), but with additional testing of supplementary doses of 0.1 and 0.01 ml/kg. Dimethylaminopropionitrile and the catalyst were also lethal to rats at the same doses and with the same symptoms. In the 0.25 ml/kg group, tremors, tonic contractions of the hind legs and loss of micturition response (on injection or when electrically stimulated, no further details) occurred during the first 3 days of dimethylaminopropionitrile treatment. A similar profile of symptoms was seen after administration of the catalyst, but with generalised malaise and polyuria on the first day of treatment. The condition and appearance of the rats returned to normal during the observation period. Histological examination revealed no treatment-related effects. Without giving further details, it was stated that the same toxic effects were observed in the 0.1 and 0.01 ml/kg dose groups, but to a lesser degree than in the 0.25 ml/kg group (Gad et al., 1979).

Groups of 5 male Sprague-Dawley rats (175 to 225 g) and ICR mice (25 to 30 g) were given oral doses of 175, 350 or 700 mg di-methylaminopropionitrile/kg body weight twice on one day and were killed 36 hours after treatment. Other groups were treated with 175 or 350 mg/kg in the same manner and killed 12, 24 and 36 hours after the second dose. Dose-dependent urine retention occurred in the rats. In the mice the most severe retention was observed at 350 mg/kg while that seen at the highest dose (700 mg/kg) was less pronounced. The bladder/body weight ratio was also determined for the mice 12, 24 and 36 hours after treatment, and was significantly increased in a dose- and time-dependent manner, indicating oedema of the bladder; this effect was observed in the rats on histological examination of the bladder, with diffuse submucosal and subserosal oedema, severe congestion of submucosal capillaries and petechial haemorrhages. In addition, perivascular infiltration was evident in the submucosal and subserosal regions. Macroscopic and histological examination of the rat kidney 48 hours after administration revealed hydronephrosis characterised by marked dilation of the renal pelvis with blunting of the renal papilla. Following twice daily oral administration of 525 mg/kg there were significant decreases in the pH value (6.6 ± 0.3; controls 7.7 ± 0.2) and the osmolality (650.5 ± 102.9; controls 1437 ± 218) of the urine. The protein content of the urine was significantly ($p<0.05$) increased at 3.5 ± 0.5 mg/dl (control 2.1 ± 0.1 mg/dl) and the urine contained significantly more ($p<0.05$) erythrocytes, leukocytes, triple phosphates and epithelial cells. Furthermore, plasma levels of urea nitrogen (6.8 ± 0.7 mmol/l, control 3.0 ± 0.1 mmol/l) and creatinine (221.3 ± 8 mmol/l, control 137.9 ± 14 mmol/l) were increased ($p<0.05$), while urinary levels were reduced (5.4 ± 1.3 mmol/24 hours, control 8.7 ± 1.8 mmol/24 hours and 172.1 ± 28 mmol/24 hours, control 317.8 ± 59 mmol/24 hours; $p<0.05$). The glutathione level in the liver, kidney and bladder and microsomal lipid peroxidation in the liver and kidney were also determined in male rats. Following a single oral dose of 525 mg dimethylaminopropionitrile/kg body weight, there was a significant ($p<0.05$) fall in the glutathione content, which was most marked 6 hours after treatment. At this point there was 5.9 ± 0.4 μmol/g in the liver (control 11.3 ± 0.8 μmol/g), 0.9 ± 0.08 μmol/g in the kidney (control 2.2 ± 0.21 μmol/g) and 1.4 ± 0.15 μmol/g in the bladder (control 2.2 ± 0.23 μmol/g). Lipid peroxidation (as μmol diene conjugate/mg phospholipid) was highest in the liver 24 hours after administration at 1.81 ± 0.10 (control 0.73 ± 0.08) and in the kidney 6 hours after administration at $2.82\pm.3$ (control 1.31 ± 0.15; $p<0.05$; Mumtaz et al., 1991b).

To study the subacute toxicity of dimethylaminopropionitrile, 5 male Sprague-Dawley rats were given 350 mg/kg body weight by gavage 5 times/week for 2 weeks. A further 5 male rats served as controls and were treated with the same volume of physiological saline. Body weight, water consumption and urine volume were recorded every 24 hours. Morphological and microscopic examinations of the liver, kidney and bladder were carried out at the end of the 2-week treatment period. Between days 6 and 12 there was a gradual fall in body weight, with a sharp fall between days 12 and 15 (from 210 g at the beginning of the study to 160 g). Water consumption was slightly but significantly higher than that in the controls during the first 7 days and thereafter fell sharply (from ca. 0.18 to 0.02 ml/g rat, control ca. 0.12 ml/g rat). The volume of urine increased during the first 8 days, slightly at first, but by the end of the study had fallen to an average of 0.03 ml/g (controls at end of the study 0.10 ml/g). Post-mortem and histological examinations again revealed congestion and oedema of the bladder with petechial bleeding, perivascular infiltration and hydronephrosis (Mumtaz et al., 1991b).

To find out whether dimethylaminopropionitrile itself or one of its potential metabolites caused the urine retention, groups of 8 rats were given 2 oral doses of 525 mg dimethylaminopropionitrile/kg body weight (vehicle unspecified) within 24 hours or equimolar doses (5.3 mmol) of β-aminopropionitrile (371 mg/kg), dimethylamine (238 mg/kg), dimethylglycine (546 mg/kg) or cyanoacetic acid (450 mg/kg). The rats were killed 48 hours after the second dose. Urinary retention was observed in all of the rats given dimethylaminopropionitrile (100%), in 4 of the 8 rats given cyanoacetic acid (50%) and in 2 of the 8 rats given dimethylamine (25%), while no urinary retention was seen after treatment with β-aminopropionitrile or dimethylglycine (Mumtaz et al., 1991b).

No treatment-related effects (urinalysis, histopathology) occurred in a preliminary inhalation study in groups of 5 male and 5 female SPF-Wistar rats exposed to concentrations of 0.01, 0.1 and 1.0 mg ca. 98% pure dimethylaminopropionitrile/l air for 6 hours/day, 5 times weekly for 2 weeks (BASF, 1989a). However, under the exposure conditions specified, the highest concentration of 1 mg/l corresponded to a daily oral dose of only ca. 180 mg/kg when converted and thus, compared with other studies involving oral administration, was too low to cause such disturbances in bladder function.

7.3   Skin and mucous membrane effects

In a preliminary study, the skin irritancy of dimethylaminopropionitrile was tested by applying 0.01 ml of the neat substance to the clipped abdominal skin of 5 albino rabbits. No macroscopically visible effects were found at the application site after 24 hours (Smyth et al., 1962).

The primary skin irritancy of neat dimethylaminopropionitrile was tested on the clipped dorsal skin of albino rabbits. No irritant effects were seen following exposures of 1 and 5 minutes, while "questionable reddening", which was no longer evident after 8 days, occurred 24 hours after a 20-hour application. Thus dimethylaminopropionitrile was not irritant to the skin in these studies (BASF, 1974).

The local irritant effects on the rabbit eye were tested in 5 rabbits at a series of different doses and concentrations. The severity of the lesions was graded on a scale from 1 to 6. The instillation of 0.02 ml neat dimethylaminopropionitrile led to damage to the eye of grade 5 or over (white discolouration of the cornea, iritis; Carpenter and Smyth, 1946).

A single dose of 50 μl neat dimethylaminopropionitrile was instilled into the conjunctival sac of one eye in albino rabbits. After one hour, slight redness and mild oedema occurred, while after 24 hours slight redness and pronounced clouding of the cornea were evident. These effects were reversible after 8 days. Dimethylaminopropionitrile was thus irritating to the eye (BASF, 1974).

7.4   Sensitisation

No information available.

7.5   Subchronic and chronic toxicity

No information available.

7.6   Genotoxicity

7.6.1 In vitro

Dimethylaminopropionitrile (98% pure) was tested for mutagenic activity in the Salmonella/microsome test (standard plate test and pre-incubation test) in strains TA 1535, TA 100, TA 1537 and TA 98 with and without metabolic activation (S9-mix from Aroclor 1254-induced rat liver). The concentrations tested ranged from 20 to 5000 μg/plate. No cytotoxicity occurred, even at the highest dose. Dimethylaminopropionitrile was not mutagenic in these studies (BASF, 1989b).

7.6.2 In vivo

No information available.

7.7   Carcinogenicity

No information available.

7.8   Reproductive toxicity

Unlike the metabolite β-aminopropionitrile, dimethylaminopropionitrile caused no embryotoxic effects in a preliminary study in which 2 female rats were given 0.3% in the diet during pregnancy (ca. 200 mg/kg body weight; no further details; Stamler, 1955).

7.9   Effects on the immune system

No information available.

7.10 Neurotoxicity

The administration of dimethylaminopropionitrile in the diet (0.35% for 56 days, equivalent to ca. 233 mg/kg body weight) to Sprague-Dawley rats of both sexes caused no recognisable skeletal alterations (osteolathyrism) and no neurological effects characteristic of a lathyrogenic substance (neurolathyrism). This is in contrast to the effects of β-aminopropionitrile (Bachhuber et al., 1955).

7.11 Other effects

The pharmacodynamic effects of dimethylaminopropionitrile and the catalyst ESN-NIAX were studied in vitro and in vivo. The in vitro studies were carried out in isolated tissue from the ileum, in nerve-muscle preparations from the diaphragm and in the muscle tissue of the bladder, with the contractile power of the musculature measured over time. Standard agonists used included carbachol (direct parasympathomimetic), neostigmine (indirect parasympathomimetic), tetramethyl ammonium chloride (ganglionic stimulant) and atropine (parasympatholytic). In addition, possible metabolites of dimethylaminopropionitrile and the catalyst were tested (incubation for 30 minutes at 37 °C with rat liver tissue, subsequent filtration of the solution and testing on the isolated organs). The catalyst initially caused stimulation of the musculature of the ileum and bladder, followed by depression which was reversible and could be overcome by higher doses of carbachol or by rinsing the preparation. Dimethylaminopropionitrile similarly showed initial stimulation of the musculature of the ileum and bladder, followed by a reduced reaction to lower doses of carbachol. The treated tissues were sensitive to high doses of carbachol. After the tissues were rinsed, there was also a reaction to low concentrations of carbachol. Neither the catalyst nor dimethylaminopropionitrile had an effect on the ganglionic diaphragm preparation. The test samples which underwent metabolic activation had no activity above that of the starting substances (Gad et al., 1979).

In the first in vivo study, the daily micturition response of 5 male Sprague-Dawley rats was studied following daily intraperitoneal injection of 0.25 ml ESN-NIAX or dimethylaminopropionitrile/kg body weight. One hour after the 5th injection, one group was given an additional 1 ml of physiological saline by intraperitoneal injection, while further groups were injected with 5 mg carbachol in 1 ml water or 0.1 ml neostigmine in 1 ml water. The urination of the rats was observed for 4 hours. Those that had been treated with carbachol or neostigmine showed lacrimation and heavy urination, while those that received only saline did not urinate during the observation period (Gad et al., 1979).

A second in vivo experiment was carried out in groups of 3 male Sprague-Dawley rats (278 to 280 g) per test substance. The direct measurement of systolic blood pressure and heart rate was facilitated by means of a cannula implanted in the abdominal aorta. The catalyst and dimethylaminopropionitrile were administered intraperitoneally, via an implanted cannula, as a series of doses (0.05, 0.15 and 0.3 ml) over a period of 2 hours. One rat per group received an additional dose of 0.5 ml of the test substance at the end of the 2-hour dosing schedule. At the end of the sequential dosing, one rat per group was given 0.5 mg atropine/kg body weight and a further rat was given 2 mg of the β-receptor blocker propanolol/kg body weight. Lacrimation, tremors and convulsions were observed in the treated rats during the 2-hour dosing period. The systolic blood pressure increased from 120–122 mmHg to 140–182 mmHg and remained elevated for about 60 minutes. The heart rate dropped from 400 beats/minute to 330 beats/minute, an effect which persisted for up to 75 minutes after administration of the substance. Arrhythmias occurred during the first 5 minutes after dosing. When atropine was injected after sequential administration of dimethylaminopropionitrile or the catalyst, the heart rate increased. After administration of propanolol, a decrease in heart rate occurred as expected. In summary, the authors established that dimethylaminopropionitrile or ESN-NIAX caused tremors, convulsions, loss of micturition reflex, increased blood pressure and loss of cardiovascular reflexes (no further details) in the rat. The lowest effective dose was said to be 0.01 ml/kg (Gad et al., 1979).

The suggestion that the production of bladder dysfunction by dimethylaminopropionitrile could involve inhibition of monoamine oxidase, was not confirmed. No inhibitory effect on monoamine oxidase activity was observed in studies with rat liver microsomes in vitro or in in vivo studies in rats involving the daily administration of 250 mg/kg body weight/day by intraperitoneal injection for 4 days (Wilmarth and Froines, 1991).

In vitro studies have shown that dimethylaminopropionitrile irreversibly inhibits the activities of the glycolytic enzymes glyceraldehyde-3-phosphate dehydrogenase and fructose-6-phosphate kinase. In contrast, the activity of lactate dehydrogenase was unaffected (Froines and Watson, 1985).

8.   Experience in humans

Following the introduction of ESN-NIAX as a catalyst in the production of plastics, it was observed that neurological dysfunctions, particularly of the bladder, were occurring in numerous workers in two American factories in which polyurethane foam was manufactured. The main symptoms were urine retention, delayed and impeded urination, weakened flow of urine and prolonged duration of micturition (Feldman et al., 1979; Kreiss et al., 1980; Landrigan et al., 1980; Landrigan, 1990; Keogh et al., 1980; Keogh, 1983; Check, 1979). The number of sick workers appeared to be correlated with the amount of catalyst used (Kreiss et al., 1980; Check, 1979).

Of 166 people who were exposed to the catalyst in one factory, 20 of 36 women (56%) and 84 or 130 men (65%), showed the above-mentioned symptoms. In addition, many patients complained of disturbances of sexual function (loss of libido, problems with erection and ejaculation), paraesthesia of the hand and feet and hyporeflexia. A comparative study of blood (activities of aspartate- and alanine aminotransferases, bilirubin, glucose, albumin, creatinine, calcium and uric acid) and urine samples (sediment) from the exposed and non-exposed subjects revealed that only the albumin content of the blood was elevated in the exposed workers (Check, 1979).

The neurological status of 8 workers (5 men and 3 women) with bladder dysfunction was investigated. Disturbances of conduction in the peripheral nerves were found with damage primarily to the sensory pathways (Feldman et al., 1979; Kreiss et al., 1980).

It is worth noting that the toxic effects also occurred in workers who had no skin contact with the catalyst, which supports uptake through the respiratory tract (Kreiss et al., 1980).

In the other affected factory, ca. 60% of the production workers who were exposed to ESN-NIAX (85 of 141) showed the symptoms of neurological dysfunction of the bladder described above and also complained of dizziness, weakness, occasional clouded vision, paraesthesia and reduced response to stimulation of the skin of the hands and feet (pricking with a needle, temperature, light contact), as well as skin effects ("dermatitis") and irritation of the respiratory tract. By means of intravenous pyelograms it was found that about 28% of the affected subjects showed decreased bladder emptying. Laboratory tests on the blood (electrolytes, urea, calcium, uric acid, phosphorus, glucose, creatinine, protein, bilirubin, cholesterol) and urine (blood, sugar, protein, ketones, bilirubin) revealed no unusual findings. After the use of the catalyst was discontinued, no further cases occurred (Keogh et al., 1980).

The so-called "DMAPN-syndrome" appears to be a largely reversible condition, as the signs of toxicity disappeared after removal of the catalyst from the production process. Thus, 86% of the affected workers at one of the factories were free of symptoms when questioned 3 months after the use of the catalyst was stopped (Kreiss et al., 1980).

After 2 years, 11 of the original dimethylaminopropionitrile-exposed sick workers were again questioned. The urological and neurological symptoms described above had largely disappeared, although there were isolated reports of difficulties in micturition and sexual dysfunction (Baker et al., 1981).

In 1982, a morbidity study was carried out in workers in a German industrial concern who had been exposed to dimethylaminopropionitrile for 9.35 (from 1.1 to 33.5) years on average. The concentrations could not subsequently be determined. No effects on health such as those described in the American publications occurred in any of the 34 workers (Deckert et al., 1982).

9.   Threshold limit values

In the Federal Republic of Germany, dimethylaminopropionitrile is classified in category II b of the list of MAK values. This contains substances for which insufficient information is available from either experience in man or from experiments in animals for a MAK value to be established (DFG, 1993).

A threshold value of 10 mg/m^3 applies in the Commonwealth of Independent States (Bayer, 1994).

References

Ahmed, A.E., Farooqui, M.Y.H.
Metabolism and toxicity of N,N-dimethylaminopropionitrile (DMAPN) in rats
Pharmacologist, 199, ref. no. 376 (1984)

Bachhuber, T.E., Lalich, J.J., Angevine, D.M., Schilling, E.D., Strong, F.M.
Lathyrus factor activity of beta-aminopropionitrile and related compounds
Proc. Soc. Exp. Biol. Med., 89, 294–297 (1955)

Baker, E.L., Christiani, D.C., Wegman, D.H., Siroky, M., Niles, C.A., Feldman, R.G.
Follow-up studies of workers with bladder neuropathy caused by exposure to dimethylaminopropionitrile
Scand. J. Work Environ. Health, 7, 54–59 (1981)

BASF AG, Medizinisch-Biologische Forschungslaboratorien
Ergebnis der gewerbetoxikologischen Vorprüfung N,N-Dimethyl-aminopropionitril
Unpublished report no. XXIV 513 (1974)

BASF AG, Abteilung Toxikologie
Kurzbericht über die toxikologische Prüfung von Dimethylamino-propionitril als Dampf an der Ratte, Inhalation über 2 Wochen
Unpublished report, project no. 36I00024/8558 (1989a)
On behalf of BG Chemie

BASF AG, Department of Toxicology
Report on the study of 3-Dimethylaminopropionitril in the Ames test (standard plate test and preincubation test with *Salmonella typhimurium*)
Unpublished report, project no. 40M0143/884312 (1989b)

BASF AG
AIDA-Grunddatensatz Propanenitrile,3-(dimethylamino)-(9CI) (1992)

Bayer AG, WV Umweltschutz und Produktsicherheit
Databank "Nationale und internationale Grenzwerte" (1994)
Carpenter, C.P., Smyth, H.F., jr.
Chemical burns of the rabbit cornea
Am. J. Ophthalmol., 29, 1363–1372 (1946)

Check, W.A.
Toxic chemical story ends happily
JAMA, 241, 2473–2474 (1979)

Deckert, H., Thiess, A.M., Mercker, H.J.
Morbiditätsstudie bei Mitarbeitern mit Exposition gegenüber Dimethylaminopropionitril (DMAPN)
Zentralbl. Arbeitsmed., 32(5), 181–184 (1982)

DFG (Deutsche Forschungsgemeinschaft)
MAK- und BAT-Werte-Liste 1993, Mitteilung 29
VCH Verlagsgesellschaft mbH, Weinheim (1993)

Feldman, R.G, Siroky, M., Niles, C.A., Kreiss, K., Wegman, D., Krane, R.
Neurotoxic dysuria due to dimethylaminopropionitrile
Neurology, 29, 560 (1979)

Froines, J.R., Postlethwait, E.M., LaFuente, E.J., Liu, W.C.V.
In vivo and in vitro release of cyanide from neurotoxic aminonitriles
J. Toxicol. Environ. Health, 16, 449–460 (1985)

Froines, J.R., Watson, A.J.
Evaluation of the inhibition of glycolytic enzymes by the neurotoxicant dimethylaminopropionitrile
J. Toxicol. Environ. Health, 16, 469–479 (1985)

Gad, S.C., McKelvey, J.A., Turney, R.A.
NIAX catalyst ESN: subchronic neuropharmacology and neurotoxicology
Drug Chem. Toxicol., 2, 223–236 (1979)

Hartung, R.
Cyanides and nitriles
in: Clayton, G.D., Clayton, F.E. (eds.)
Patty's industrial hygiene and toxicology
3rd ed., vol. 2C, p. 4881–4882
John Wiley and Sons, New York, Chichester, Brisbane, Toronto, Singapore (1982)

Jaeger, R.J., Plugge, H., Szabo, S.
Acute urinary bladder toxicity of a polyurethane foam catalyst mixture: a possible new target organ for a propionitrile derivative
J. Environ. Pathol. Toxicol., 4, 555–562 (1980)

Keogh, J.P., Pestronk, A., Wertheimer, D., Moreland, R.
An epidemic of urinary retention caused by dimethylaminopropi-
onitrile
JAMA, 243, 746–749 (1980)

Keogh, J.P.
Classical syndromes in occupational medicine: dimethylamino-
propionitrile
Am. J. Ind. Med., 4, 479–484 (1983)

Kreiss, K., Wegman, D.H., Niles, C.A., Siroky, M.B., Krane, R.J.,
Feldman, R.G.
Neurological dysfunction of the bladder in workers exposed to dim-
ethylaminopropionitrile
JAMA, 243, 741–745 (1980)

Landrigan, P.J., Kreiss, K., Xintaras, C., Feldman, R.G., Heath, C.W., jr.
Clinical epidemiology of occupational neurotoxic disease
Neurobehav. Toxicol., 2, 43–48 (1980)

Landrigan, P.J.
Prevention of toxic environmental illness in the twenty-first century
Environ. Health Perspect., 86, 197–199 (1990)

Mumtaz, M.M., Farooqui, M.Y.H., Ghanayem, B.I., Ahmed, A.E.
The urotoxic effects of N,N′-dimethylaminopropionitrile. 2. In vivo
and in vitro metabolism
Toxicol. Appl. Pharmacol., 110, 61–69 (1991a)

Mumtaz, M.M., Farooqui, M.Y.H., Ghanayem, B.I., Rajaraman, S.,
Frankenberg, L., Ahmed, A.E.
Studies on the mechanism of urotoxic effects of N,N′-dimethyl-
aminopropionitrile in rats and mice. 1. Biochemical and morpholo-
gic characterisation of the injury and its relationship to metabolism
J. Toxicol. Environ. Health, 33, 1–17 (1991b)

NIOSH (National Institute for Occupational Safety and Health)
NIAX Catalyst ESN
Current Intelligence Bulletin, 26 (1978)

Smyth, H.F., jr., Carpenter, C.P., Weil, C.S., Pozzani, U.C., Striegel, J.A.
Range-finding toxicity data: list VI
Am. Ind. Hyg. Assoc. J., 23, 95–107 (1962)

Stamler, F.W.
Reproduction in rats fed Lathyrus peas or aminonitriles
Proc. Soc. Exp. Biol. Med., 90, 294–298 (1955)

UCC (Union Carbide Corporation)
Product information note (1978)
Cited in: Jaeger et al. (1980)

Vitorskii, A.P.
Toxicology of nitriles
Chem. Abstr. 57, 25361 (1961)
Cited in: Jaeger et al. (1980)

Wilmarth, K.R., Froines, J.R.
Role of monoamine oxidase in aminopropionitrile-induced neuro-
toxicity
J. Toxicol. Environ. Health, 32, 415–427 (1991)

Ethenesulfonic acid, sodium salt

1. Summary and assessment

Based on the available studies in animals, ethenesulfonic acid, sodium salt is of low acute toxicity (LD_{50} rat oral >15,000 mg/kg body weight).

A 30% aqueous solution of ethenesulfonic acid, sodium salt is not irritating to the skin or eyes of the rabbit.

The substance is not a skin sensitiser in the Magnusson and Kligman maximisation test.

Ethenesulfonic acid, sodium salt is not mutagenic in the Salmonella/microsome test.

The sodium salt of ethenesulfonic acid is being processed within the OECD "High Production Volume Program" which has been assigned to Belgium. The OECD has requested that a dermal LD_{50} test, a chromosome aberration test and a "combined test" (combined repeated dose toxicity study with reproduction/developmental toxicity screening test) should be carried out.

2. Name of substance

2.1	Usual name	Ethenesulfonic acid, sodium salt
2.2	IUPAC-name	Ethenesulfonic acid, sodium salt
2.3	CAS-No.	3039–83–6
2.4	EINECS-No.	221–242–5

3.   Synonyms, common and trade names

Ethensulfonsäure, Natriumsalz
Sodium ethenesulfonate
Sodium ethylenesulfonate
Sodium vinylsulfonate
Tamol VS
Vinylsulfonic acid sodium

4.   Structural and molecular formulae

4.1   Structural formula

4.2   Molecular formula $C_2H_3O_3SNa$

5.   Physical and chemical properties

5.1   Molecular mass, g/mol   130.1

5.2   Melting point, °C   < –20 (30% aqueous solution)
(Hoechst, 1990)

5.3   Boiling point, °C   No information available

5.4   Vapour pressure, hPa   No information available

5.5   Density, g/cm^3 1.2 (30% aqueous solution)
(Hoechst, 1988a)

5.6   Solubility in water   Completely miscible
(Hoechst, 1988a)

5.7   Solubility in
organic solvents   No information available

5.8   Solubility in fat   Insoluble (Hoechst, 1988a)

5.9   pH value   11.5–12 (30%)

$\qquad\qquad\qquad\qquad\qquad\qquad\qquad\qquad$ (Hoechst, 1988a)

5.10 Conversion factor   1 ml/m³ (ppm) $\cong$ 5.31 mg/m³
$\qquad\qquad\qquad\qquad\qquad\qquad$ 1 mg/m³ $\cong$ 0.19 ml/m³ (ppm)
$\qquad\qquad\qquad\qquad\qquad\qquad$ (at 1013 hPa and 25 °C)

6.   Uses

As a co-monomer and in the manufacture of dispersions; intermediate in organic synthesis (Hoechst, 1988c).

7.   Experimental results

7.1   Toxicokinetics and metabolism

No information available.

7.2   Acute and subacute toxicity

The LD_{50} value in female rats following a single oral dose was >15,000 mg/kg body weight. Signs of toxicity included prostration, and the animals had unkempt coats. No macroscopic effects were evident in the rats at the end of the 14-day observation period (Hoechst, 1976a).

7.3   Skin and mucous membrane effects

In a patch test, 0.5 ml of a 30% aqueous solution of ethenesulfonic acid, sodium salt was applied to the clipped, intact or scarified skin of 6 rabbits (White Russian) for 24 hours under occlusive cover. The results were recorded immediately after removal of the patch, and 48 and 72 hours after the beginning of the study. Only one rabbit showed very slight oedema of both the intact and scarified skin, at both 24 and 48 hours after the beginning of the study, an effect which had cleared up after 72 hours. An irritation index of 0.08 out of a maximum of 8 was determined (Hoechst, 1976b). The 30% aqueous solution was thus not irritating.

Pairs of female Pirbright white guinea pigs were treated with 0.5 ml of a 1, 10 or 30% aqueous solution of ethenesulfonic acid, sodium salt, which was applied to the clipped flank skin in a further patch test. The duration of exposure (occlusive) was 24 hours. With the 30% solution, clear, localised erythema and pustule formation, as well as barely perceptible oedema, were evident at the application site 48 hours after the beginning of treatment. The 1 and 10% solutions did not give rise to any irritant effects (Hoechst, 1988a).

In an eye irritation study, 0.1 ml of a 30% aqueous solution of ethenesulfonic acid, sodium salt was instilled into the conjunctival sac of one eye of each of 6 rabbits (white Russian), the second eye serving as an untreated control. The study was carried out in accordance with FDA guidelines. No irritant effects were detected during a 72-hour observation period (Hoechst, 1976b).

7.4   Sensitisation

A 30% aqueous solution of ethenesulfonic acid, sodium salt was tested for its ability to induce skin sensitisation in 10 Pirbright white guinea pigs in a maximisation test according to Magnusson and Kligman. A further 5 animals were used as controls. For induction, the animals received intradermal injections, into the shoulder region, of 5% preparations (0.1 ml) of the substance in Freund's adjuvant and in 0.9% saline; 8 days later they were treated with 0.5 ml of a 10% solution of ethenesulfonic acid, sodium salt in 0.9% saline, which was applied percutaneously for 48 hours under occlusive cover to the area of the intradermal injection site. The concentrations used led to minimal irritation at the application site. The challenge treatment was carried out with 0.5 ml of a 10% solution of ethenesulfonic acid, sodium salt in 0.9% saline, which was applied to the left flank 14 days after the second induction treatment. There were no indications of a sensitising effect (Hoechst, 1988a).

7.5   Subchronic and chronic toxicity

No information available.

7.6   Genotoxicity

7.6.1 In vitro

A 30% aqueous solution of ethenesulfonic acid, sodium salt was tested for mutagenic activity in the *Salmonella typhimurium* strains TA 98, TA 100, TA 1535, TA 1537 and TA 1538 and in *Escherichia coli* WP2uvrA with and without metabolic activation (S9-mix from Aroclor 1254-induced rat liver) in the direct plate test. The concentrations tested ranged from 4 to 10,000 µg/plate and 3 plates were used at each concentration. The test was repeated in an independent experiment. There were no indications of mutagenic activity (Hoechst, 1988b).

7.6.2 In vivo

No information available.

7.7   Carcinogenicity

No information available.

7.8   Reproductive toxicity

No information available.

7.9   Effects on the immune system

No information available.

7.10 Neurotoxicity

No information available.

7.11 Other effects

No information available.

8.   Experience in humans

No information available.

9.   Threshold limit values

No information available.

References

Hoechst AG, Pharma Forschung Toxikologie
Akute orale Toxizität von Vinylsulfonsaurem Natrium (=Natriumvi-
nylsulfonat) an weiblichen SPF-Wistar-Ratten
Unpublished report no. 42/76 (1976a)

Hoechst AG, Pharma Forschung Toxikologie
Haut- und Schleimhautverträglichkeit von Vinylsulfonsaurem
Natrium (=Natriumvinylsulfonat) an Kaninchen
Unpublished report no. 43/76 (1976b)

Hoechst AG, Pharma Forschung Toxikologie und Pathologie
Vinylsulfonsaures Natrium, 30%ig - Prüfung auf sensibilisierende
Eigenschaften an Pirbright-White-Meerschweinchen im Maximie-
rungstest
Unpublished report no. 88.0446 (1988a)

Hoechst AG, Pharma Research Toxicology and Pathology
Vinylsulfonsaures Natrium (30%) - Study of the mutagenic potential
in strains of *Salmonella typhimurium* (Ames test) and *Escherichia
coli*
Unpublished report no. 88.0444 (1988b)

Hoechst AG, Abteilung UCV
Data sheet "Altstoffe" Ethensulfonsäure, Natriumsalz (1988c)

Hoechst AG
DIN-Safety data sheet "Vinylsulfonsaures Natrium 30%" (1990)
SIDS Profile (Screening Information Data Set)
Ethenesulfonic acid, sodium salt
OECD High Production Volume Chemicals Program (undated)

N,N-Dicyclohexyl-2-benzothiazolesulfenamide

1.   Summary and assessment

The acute oral and dermal toxicity of N,N-dicyclohexyl-2-benzo-thiazolesulfenamide is low (LD_{50} rat oral > 5000 mg/kg body weight; LD_{50} rabbit dermal >2000 mg/kg body weight).

N,N-Dicyclohexyl-2-benzothiazolesulfenamide is not irritating to the skin and slightly irritating to the eye in the rabbit.

The substance did not induce sensitisation in the maximisation test in the guinea pig.

Subacute and subchronic administration in the diet of rats led only to reduced body weight gain associated with reduced feed consumption. There were no effects on clinical chemistry or hae-matological parameters and nothing abnormal was detected on pathological and histopathological examination. The no effect lev-el on subchronic administration in the diet was 500 mg/kg diet. Irri-tant effects on the mucous membranes occurred in rats after repeat-ed exposure by inhalation (0.35 to 0.40 mg/l, 2 hours/day for 15 days). No treatment-related systemic effects were detected.

There are no indications that N,N-dicyclohexyl-2-benzothiazole-sulfenamide has genotoxic properties. It is not mutagenic in the Sal-monella/microsome test or the HPRT test either with or without metabolic activation, it does not cause DNA damage in rat hepato-cytes in the UDS test and does not induce chromosome damage in the chromosome aberration test in the rat.

An increased incidence of sarcomas at the site of application occurred in a preliminary carcinogenicity study in the mouse, involving the subcutaneous application of a total of 20,000 mg N,N-dicyclohexyl-2-benzothiazolesulfenamide/kg body weight, as single doses of 1000 mg/kg body weight applied at irregular inter-vals over 413 days. The authors of the study judged this to be an indication that N,N-dicyclohexyl-2-benzothiazolesulfenamide had a weak carcinogenic effect, but pointed out that the subcutaneous tissues of the rat are very sensitive to tumour formation, and thus tumours at the injection site are difficult to evaluate in terms of the possible carcinogenic potential of a substance. The product has no genotoxic potential based on tests with various genetic end points, and no effects on the tissues have been observed in studies involving repeated administration. For these reasons, it is unlikely that the substance has carcinogenic potential.

2.   Name of substance

2.1   Usual name   N,N-Dicyclohexyl-2-benzo-
                           thiazolesulfenamide

2.2   IUPAC-name   N,N-Dicyclohexyl-2-benzo-
                           thiazolesulfenamide

2.3   CAS-No.   4979–32–2

2.4   EINECS-No.   225–625–8

3.   Synonyms, common and trade names

2-Benzothiazolesulfenamide,
  N,N-dicyclohexyl
2-Benzothiazolyl-N,N-dicyclo-
  hexylsulfenamide
DCBS
Ekalund DCBS
N,N-Dicyclohexylbenzothiazo-
  lesulfenamide
N,N-Dicyclohexyl-2-
  benzothiazolsulfenamide
N,N-Dicyclohexyl-2-
  benzothiazolylsulfenamide
Perkacit DCBS
Rhodifax 30

Santocure DCBS
Sulfenamid DC
Vulkacit DZ

4.   Structural and molecular formulae

4.1   Structural formula

4.2   Molecular formula $C_{19}H_{26}N_2S_2$

5.   Physical and chemical properties

5.1   Molecular mass, g/mol   346.6

5.2   Melting point, °C   >96   (Bayer, 1992)

5.3   Boiling point, °C   –

5.4   Vapour pressure, hPa   ca. 0.00075 (at 120 °C)
        (Bayer, 1992)
        ca. 0.03 (at 140 °C)
        (Bayer, 1992)

5.5   Density, g/cm³   1.2 (bulk density) (Bayer, 1992)

5.6   Solubility in water   0.03 g/l (at 25 °C) (Bayer, 1992)

5.7   Solubility in
organic solvents   Soluble in ethanol and most
        organic solvents (WTR, 1984)

5.8   Solubility in fat   No information available

5.9   pH value   No information available

5.10 Conversion factor

$1 \text{ ml/m}^3 \text{ (ppm)} \triangleq 14.15 \text{ mg/m}^3$
$1 \text{ mg/m}^3 \triangleq 0.07 \text{ ml/m}^3 \text{ (ppm)}$
(at 1013 hPa and 25 °C)

6.   Uses

Vulcanisation accelerator in the rubber industry (Bayer, 1992).

7.   Experimental results

7.1   Toxicokinetics and metabolism

No information available.

7.2   Acute and subacute toxicity

The following LD_{50} values were determined following acute exposure to N,N-dicyclohexyl-2-benzothiazolesulfenamide:

Species	Route	LD_{50} (mg/kg body weight)	Reference
Rat	oral	6420	Marhold, 1986
Rat	oral	10000	TNO, 1975
Rat	oral	>5000	Monsanto, 1991
Mouse	oral	8500	Vorob'eva, 1968
Rabbit	dermal	>2000	Monsanto, 1991
Rat	s.c.*	>5000	Bayer, 1975

* Technically and analytically pure substance

Signs of toxicity seen in the rat following acute oral administration included reduced mobility and discharges from the eyes and nose. No pathological effects were evident on macroscopic examination. The animals died between 3 and 8 days after administration of the substance (TNO, 1975). Slight necrotic effects on the gastric mucosa were described in the mouse following oral administration (no further details; Vorob'eva, 1968).

Groups of 5 male and 5 female Sprague-Dawley rats were given N,N-dicyclohexyl-2-benzothiazolesulfenamide in their diet at concentrations of 0, 2000, 3000, 5000, 7500 and 10,000 mg/kg feed (equivalent to ca. 133, 200, 333, 500 and 666 mg/kg body weight) for 4 weeks. The administration of N,N-dicyclohexyl-2-benzothiazolesulfenamide led to a dose-dependent reduction in body weight gain with reduced feed intake. Post-mortems revealed no treatment-related pathological effects; the clinical chemistry and haematological parameters and the absolute and relative organ weights were unchanged when compared with controls (no further details; Monsanto, 1991).

Male rats were exposed to 350 to 400 mg N,N-dicyclohexyl-2-benzothiazolesulfenamide/m^3 air for 2 hours/day for 15 days. Other than irritation of the mucous membranes, no overt signs of toxicity were observed. There were no effects on the liver or the kidney (no further details; Vorob'eva, 1968).

7.3   Skin and mucous membrane effects

N,N-Dicyclohexyl-2-benzothiazolesulfenamide was described as not irritating to the skin or eyes in a summary document (no further details; WTR, 1984).

N,N-Dicyclohexyl-2-benzothiazolesulfenamide was not irritating to the skin of the rabbit (no further details; Monsanto, 1991).

In a further study, moderate irritant effects were found in rabbits following dermal application of 20 mg N,N-dicyclohexyl-2-benzothiazolesulfenamide for 24 hours (no further details; Marhold, 1986).

Mild irritation of the conjunctiva was seen 1 hour after the introduction of N,N-dicyclohexyl-2-benzothiazolesulfenamide into the rabbit eye. No abnormal effects were detected in the rabbits 24 hours after introduction. The compound was evaluated as practically non-irritant to the rabbit eye (no further details; Monsanto, 1991).

Mild irritant effects on the eye of the rabbit were reported following the introduction of 500 mg N,N-dicyclohexyl-2-benzothiazolesulfenamide (no further details; Marhold, 1986).

7.4   Sensitisation

N,N-Dicyclohexyl-2-benzothiazolesulfenamide did not induce sensitisation in a maximisation test in guinea pigs (no further details; Monsanto, 1991).

7.5   Subchronic and chronic toxicity

The administration of N,N-dicyclohexyl-2-benzothiazolesulfen-amide to rats at concentrations of 2500 and 5000 mg/kg feed (equivalent to ca. 166 and 333 mg/kg body weight) for 3 months caused reductions in body weight gain and feed consumption in both males and females. There were no histopathological effects and no indications of target organs. The no effect level was given as 500 mg N,N-dicyclohexyl-2-benzothiazolesulfenamide/kg feed (equivalent to ca. 33 mg/kg body weight; no further details; Monsanto, 1991).

7.6   Genotoxicity

7.6.1 In vitro

N,N-Dicyclohexyl-2-benzothiazolesulfenamide was tested for mutagenic properties in the direct plate test in the *Salmonella typhimurium* strains TA 98 and TA 100 at concentrations ranging from 1 to 1000 μg/plate with and without metabolic activation (S9-mix from PCB-induced rat liver). At each concentration and for each strain, 3 plates were used. The study was repeated in an independent experiment. N,N-Dicyclohexyl-2-benzothiazolesulfenamide showed no mutagenic effects (You et al., 1982).

In a further Salmonella/microsome test, N,N-dicyclohexyl-2-benzothiazolesulfenamide gave no indication of mutagenic activity either with or without metabolic activation (no further details; Monsanto, 1991).

N,N-Dicyclohexyl-2-benzothiazolesulfenamide did not cause gene mutations in the HPRT test in CHO cells. Concentrations of up to 500 μg/ml were tested, both with and without metabolic activation (no further details; Monsanto, 1991).

N,N-Dicyclohexyl-2-benzothiazolesulfenamide did not cause any DNA damage in the UDS test in rat hepatocytes at concentrations of up to 50 μg/ml (no further details; Monsanto, 1991).

7.6.2 In vivo

In a test for chromosome aberrations in the rat, the oral administration of 1000 mg N,N-dicyclohexyl-2-benzothiazolesulfenamide/kg body weight did not give rise to any chromosomal damage in the bone marrow (no further details; Monsanto, 1991).

7.7   Carcinogenicity

In a preliminary carcinogenicity study, 100-day-old Wistar rats were given Vulkacit DZ, chemically pure (99% N,N-dicyclohexyl-2-benzothiazolesulfenamide, 0.1% dicyclohexylamine, 0.9% unidentified residue) or Vulkacit DZ, technically pure (ca. 98.6% N,N-dicyclohexyl-2-benzothiazolesulfenamide, 0.2% dicyclohexylamine, ca. 1.2% unidentified residue) once a week or at greater intervals (depending on the condition of the rats; no detailed information given). The test substances were administered by subcutaneous injection in the middle of the back at doses of 1000 mg/kg body weight in each case. The total dose was 20,000 mg/kg body weight. Finely powdered Vulkacit DZ was moistened with ethanol and administered as a suspension in physiological saline, the controls receiving a 10% solution of ethanol in physiological saline. The treated groups consisted of 20 rats per sex, while the control group contained 25 per sex. The final injection was administered after 413 days of the study and the rats were observed until their natural deaths. They were then examined macroscopically and histopathologically. No conspicuous treatment-related effects were seen on the behaviour, appearance, weight gain or survival of the animals. Treatment with N,N-dicyclohexyl-2-benzothiazolesulfenamide led to an increased incidence of local sarcomas at the injection site (28 and 33%, respectively, controls 2%). The incidences of other tumours were within the range of control and historical control values. The authors pointed out that the subcutaneous tissues of the rat are highly susceptible to tumour formation, making tumours at the injection site difficult to evaluate in connection with the estimation of the possible carcinogenicity of a substance. They judged the increased incidence of local sarcomas at the injection site to be an indication that N,N-dicyclohexyl-2-benzothiazolesulfenamide had a weak carcinogenic effect (Bayer, 1975).

According to tests with various genetic endpoints, the product has no genotoxic potential, and studies involving repeated administration have produced no effects on the tissues. For these reasons, it is unlikely that the substance has carcinogenic potential.

7.8   Reproductive toxicity

N,N-Dicyclohexyl-2-benzothiazolesulfenamide was not toxic or teratogenic at concentrations of up to 0.5 μmol per egg (equivalent to 173 μg/egg) when introduced into the air chamber of 3-day-old hen's eggs (volume administered 5 μl). At least 20 eggs were used per concentration (Korhonen et al., 1982, 1983). The extrapolation of these findings to mammalian systems is, however, not really appropriate, as the absorption barriers and detoxification mechanisms present in mammals are not represented in this system. Furthermore, this test system is over-sensitive and cannot distinguish between teratogenic and embryolethal effects (Skofitsch, 1988; Neubert et al., 1992; Neubert, 1993; Heinrich-Hirsch, 1992).

7.9   Effects on the immune system

No information available.

7.10   Neurotoxicity

No information available.

7.11   Other effects

No information available.

8.   Experience in humans

No information available.

9.   Threshold limit values

No information available.

References

Bayer AG, Institut für Toxikologie
Vulkacit DZ (technisch rein und chemisch rein) – Orientierende
Cancerogenese-Versuche mit subkutaner Gabe an Ratten
Unpublished report no. 5119 (1975)

Bayer AG
AIDA-Grunddatensatz 2-benzothiazolesulfenamide, N,N-Dicyclo-
hexyl- (1992)

Heinrich-Hirsch, B.
Possible contribution of in vitro methods risk assessment; discus-
sion of the presentation
In: Neubert et al., p. 471 (1992)

Korhonen, A., Hemminki, K., Vainio, H.
Embryotoxicity of benzothiazoles, benzenesulfohydrazide, and
dithiodimorpholine to the chicken embryo
Arch. Environ. Contam. Toxicol., 11, 753–759 (1982)

Korhonen, A., Hemminki, K., Vainio, H.
Toxicity of rubber chemicals towards three-day chicken embryos
Scand. J. Work Environ. Health, 9, 115–119 (1983)

Marhold, J.V.
Prehled Prumyslove Toxicol. Org. Latky, p. 1101 (1986)
Cited in: RTECS (1990)

Monsanto Services International S.A./N.V.
Written communication to BG Chemie of 30 September 1991

Neubert, D., Kavlock, R.J., Merker, H.J., Klein, J. (eds.)
Risk assessment of prenatally-induced adverse health effects
Springer-Verlag (1992)

Neubert, D.
Personal communication (1993)

RTECS (Registry of Toxic Effects of Chemical Substances)
2-Benzothiazolesulfenamide, N,N-dicyclohexyl-, DL6300000
Produced by NIOSH (National Institute for Occupational Safety
and Health) (1990)

Skofitsch, G.
Vom Hund zum Ei? Große und kleine Laboratoriumstiere
In: Lembeck, F. (ed.)
Alternativen zum Tierversuch, p. 71–79
Georg Thieme Verlag Stuttgart, New York (1988)

TNO, Centraal Instituut voor Voedingsonderzoek
Determination of the acute oral toxicity of Rhodifax 30 in rats
Unpublished report (1975)
On behalf of the Working Group on Toxicology of Rubber Auxiliar-
ies (WTR)

Vorob'eva, R.S.
N,N-Dicyclohexyl-2-benzothiazolesulfenamide
Toksikol. Nov. Khim. Veshchestv. Vnedryaemykh Rezin. Shinnuyu
Prom., 89–93 (1968)
Cited in: Chem. Abstr., 71, 176, 20566 h (1969)

WTR (International Working Group on the Toxicology of Rubber
Additives)
Safety data sheet N,N-dicyclohexyl-2-benzothiazolesulphenamide
(1984)

You, X., Zhou, Y., Hu, Y.
Mutagenitätsuntersuchungen an 14 Beschleunigern für Gummi
(Brief German translation of the Chinese)
Huanjing Kexue, 3, 39–42 (1982)

2,5-Dimethoxy-4-chloroaniline

1. Summary and assessment

2,5-Dimethoxy-4-chloroaniline is of low to moderate acute toxicity to rats on oral and dermal exposure (LD_{50} rat oral 1260 and 4800 mg/kg body weight, rat dermal >2000 mg/kg body weight). On oral administration to rats in a 28-day subacute study, 2,5-dimethoxy-4-chloroaniline caused anaemia and liver and kidney damage. The no toxic effect level was 20 mg/kg body weight.

The substance is not irritating to the dorsal skin or the eye of the rabbit.

2,5-Dimethoxy-4-chloroaniline is mutagenic following metabolic activation in the Salmonella/microsome test, but is negative in the HPRT test in Chinese hamster V79 cells and is not genotoxic in the micronucleus test in mice.

2. Name of substance

2.1	Usual name	2,5-Dimethoxy-4-chloroaniline
2.2	IUPAC-name	2,5-Dimethoxy-4-chloroaniline
2.3	CAS-No.	6358–64–1
2.4	EINECS-No.	228–782–0

3. Synonyms, common and trade names

1-Amino-4-chloro-2,5-dimethoxybenzene

2-Amino-5-chloro-hydro-
quinone dimethylether
Chloroaminohydroquinone
dimethylether
4-Chloro-2,5-dimethoxy
benzenamine
CMEB
CME-Base
2,5-Dimethoxy-4-chloranilin

4.   Structural and molecular formulae

4.1   Structural formula

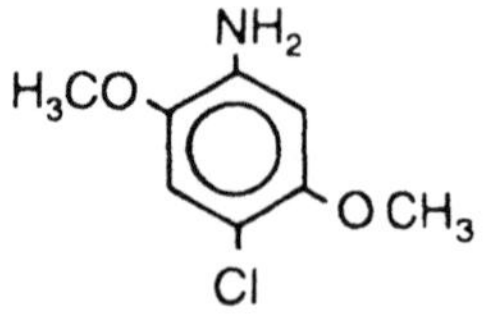

H_3CO—(ring with NH_2 top, OCH_3 and Cl)—

4.2   Molecular formula $C_8H_{10}ClNO_2$

5.   Physical and chemical properties

5.1   Molecular mass, g/mol   187.63

5.2   Melting point, °C   117.5   (Hoechst, 1991a)

5.3   Boiling point, °C   No information available

5.4   Vapour pressure, hPa   0.00001 (at 20 °C)
                (Hoechst, 1991a)
        0.0933 (at 90 °C)
                (Hoechst, 1991a)

5.5   Density, g/cm³   1.3 (at 20 °C) (Hoechst, 1986a, b)

5.6   Solubility in water   4 g/l (at 20 °C)   (Hoechst, 1991a)

5.7	Solubility in organic solvents	Soluble in ethanol, toluene, acetone, xylene, benzene, ether (Hoechst, 1986a, b)
5.8	Solubility in fat	No information available
5.9	pH value	7 (Hoechst, 1986c)
5.10	Conversion factor	1 ml/m^3 (ppm) $\cong$ 7.66 mg/m^3 1 mg/m^3 $\cong$ 0.13 ml/m^3 (ppm) (at 1013 hPa and 25 °C)

6. Uses

Intermediate in the production of dyestuffs (Hoechst, 1986a, b).

7. Experimental results

7.1 Toxicokinetics and metabolism

No information available.

7.2 Acute and subacute toxicity

The acute oral toxicity of 2,5-dimethoxy-4-chloroaniline was studied in groups of 10 female Wistar rats (180 to 196 g), which were treated with 800, 2000 or 5000 mg/kg body weight. The observation period was 14 days. An LD_{50} value of 1260 (798 to 1850) mg/kg body weight was determined. Signs of toxicity observed included stupefaction, disturbances of balance, the adoption of crouching, cringing and hunched postures, lying on the abdomen or side, ruffled fur, partly-closed eyes, increased lacrimation, irregular or spasmodic and noisy breathing and narcosis. Macroscopic effects found at post-mortem in the animals that died included marks on the lobes of the liver and bladders full of brown urine (Hoechst, 1979).

The oral LD_{50} determined in a total of 50 male Wistar rats was 4800 (3800 to 6000) mg/kg body weight after a 14-day observation period. The signs of toxicity were non-specific. No post-mortems were carried out (MB, 1979).

The dermal LD_{50} was determined in 5 male and 5 female Wistar rats in accordance with OECD guideline no. 402, and was found to be more than 2000 mg/kg body weight for both sexes. No deaths or signs of toxicity were seen and no gross pathological effects were evident at autopsy (Hoechst, 1988a).

In a preliminary 3-day feeding study in 5 wild mice (*Peromyscus maniculatus*, natural habitat North America), each animal was given 25 wheat seeds/day which had been treated with 2% of the test substance. The mice were observed for 4 days. Compared with the control group, 43% fewer wheat seeds were consumed. More than 50% of the mice died within the observation period, having eaten on average 713 mg of the substance/kg body weight/day (no further details; Schafer and Bowles, 1985).

In a preliminary study carried out prior to a 4-week feeding study, groups of 5 male and 5 female Sprague-Dawley rats (Crl:CD; average initial weights ca. 163 and 145 g, respectively) received 0, 2000, 4000 or 8000 ppm 2,5-dimethoxy-4-chloroaniline in the diet for 14 days. The study was carried out in accordance with OECD guideline no. 407. Based on feed consumption, the average daily intakes were 0, 240, 469 and 915 mg/kg body weight in the males and 0, 225, 460 and 906 mg/kg body weight in the females. No deaths occurred, but feed consumption and body weight gain were dose-dependently reduced in the first week. No substantial differences were found compared with the controls during the second week. Hair loss from the back and shoulders occurred in all treatment groups. Haematological effects in the rats in the top dose group included reductions in the haemoglobin level, erythrocyte count and haematocrit value, polychromasia and, in the female rats, increased thrombocyte counts and delayed blood clotting. Some of these effects were also seen in rats in the 4000 and 2000 ppm groups, although to a lesser extent. Clinical chemistry tests revealed increased cholesterol levels, reduced aspartate aminotransferase (glutamate-oxalate transaminase) and alkaline phosphatase activities and increased bilirubin levels in the

8000 ppm group. Increased cholesterol levels and a reduction in alkaline phosphatase activity were also seen at 4000 ppm. In the 2000 ppm group, only the alkaline phosphatase activity was reduced. All concentrations caused dose-dependent increases in liver weights and relative kidney weights. The adrenal gland weights were significantly reduced in the females in the 8000 and 4000 ppm groups. Histological examination revealed minimal centrilobular enlargement of the hepatocytes in all treatment groups. A no effect level could not be determined. Analytical studies revealed that low 2,5-dimethoxy-4-chloroaniline concentrations (50 ppm) are not stable in the diet (HRC, 1993a).

Because of the instability of low concentrations of 2,5-dimethoxy-4-chloroaniline in the diet, a 4-week study involving daily gavaging was subsequently carried out in accordance with OECD guideline no. 407, but with histological examination of the liver, kidneys and testes extended to all dose groups. Groups of 5 male and 5 female Sprague-Dawley rats (Crl:CD; average initial weights 147.2 and 137 g, respectively) were given (77.9% pure) 2,5-dimethoxy-4-chloroaniline daily as a suspension in wheatgerm oil for a period of 29 days. The doses used were 0, 20, 100 and 500 mg/kg body weight. No deaths occurred. Rats in the 500 mg/kg group showed piloerection, hunched posture, increased salivation, occasional waddling gait and lethargy. The 100 mg/kg dose also caused increased salivation, waddling gait and piloerection, but not so frequently, while at 20 mg/kg, only occasional increases in salivation were seen after gavaging. Body weight was significantly reduced in the female rats in the two higher dose groups at the end of the study. Haematological effects at the top dose level included signs of anaemia with reticulocytosis, which frequently occurred to a lesser extent in the rats in the intermediate dose group. In addition, slight polychromasia occurred in all rats in the top dose group and the majority in the intermediate dose group, with slight but significantly reduced clotting times seen in the females. Clinical chemistry tests revealed increased bilirubin levels in the two upper dose groups with a slight decrease in aspartate aminotransferase (glutamate-oxalate transaminase) activity in the males. Cholesterol levels were increased in the 500 mg/kg group, with increased plasma protein and creatinine levels seen in the females. A significant increase in early and late normoblasts was reported in the bone marrow of the males at 500 mg/kg. The absolute and relative liver weights and the relative kidney weights were increased in the top dose group, as were the absolute and relative liver weights in the intermediate dose group with the relative kidney weights being increased only in the females at this dose level. The absolute and relative spleen weights were also increased in the high dose group. The absolute testis weights were decreased in all treatment groups, although not dose-dependently. Histological examination revealed kidney damage (papillary necrosis) and centrilobular hepatocyte enlargement in the liver at 500 mg/kg body weight, effects which also occurred, although less frequently, in the rats in the 100 mg/kg group. There were no histological effects that correlated with the reduced testis weights in any of the dose groups. The no toxic effect level was 20 mg/kg body weight (HRC, 1993b).

7.3   Skin and mucous membrane effects

The skin irritancy of 2,5-dimethoxy-4-chloroaniline was tested in accordance with OECD guideline no. 404, in a study in which 500 mg of the substance was applied to the shaved skin of 3 rabbits under semi-occlusive cover for 4 hours. The skin was examined for possible irritant effects 30 and 60 minutes and 24, 48 and 72 hours after the end of exposure. No irritant effects were observed (Hoechst, 1986a).

In order to study eye irritation, 100 mg of the substance was introduced into the conjunctival sac of the left eye of each of 3 rabbits (albino, New Zealand) in accordance with OECD guideline no. 405. The irritant effect was evaluated 1, 24, 48 and 72 hours after the substance was placed in the eye. 2,5-Dimethoxy-4-chloroaniline caused minimal irritation, which was reversible after 72 hours (Hoechst, 1986b). According to the OECD classification criteria, the substance was therefore not irritating to the eye.

7.4   Sensitisation

No information available.

7.5   Subchronic and chronic toxicity

No information available.

7.6   Genotoxicity

7.6.1 In vitro

The genotoxic effect of 2,5-dimethoxy-4-chloroaniline (99.9% pure) was studied in the Salmonella/microsome test in strains TA 100, TA 1535, TA 1537, TA 1538 and TA 98 and in *Escherichia coli* WP2uvrA with and without metabolic activation (S9-mix from Aroclor 1254-induced rat liver). The substance was cytotoxic to most of the bacterial strains at concentrations of 10,000 µg/plate and above. With metabolic activation, 2,5-dimethoxy-4-chloro-aniline showed concentration-dependent mutagenicity in strains TA 100 and TA 98 and possibly also in TA 1538 (Hoechst, 1986c).

2,5-Dimethoxy-4-chloroaniline (ca. 99% pure) was also tested for genotoxic effects in Chinese hamster V79 cells in vitro. The study was carried out with and without metabolic activation (S9-mix from Aroclor 1254-induced rat liver). In preliminary studies, concentrations of 300 µg/ml (limit of solubility) were not cytotoxic without activation, but with activation a significant cytotoxic effect occurred even at 2 µg/ml. In the main study, therefore, the concentrations tested were 37.5, 75, 150 and 300 µg/ml without metabolic activation and 0.25, 0.5, 1.0, 1.25 and 1.5 µg/ml with activation. 2,5-Dimethoxy-4-chloroaniline was not mutagenic under these study conditions and caused no relevant cytotoxic effects (Hoechst, 1988b).

7.6.2 In vivo

2,5-Dimethoxy-4-chloroaniline (77.3% pure) was studied in the micronucleus test in accordance with OECD guideline no. 474. Groups of 5 male and 5 female NMRI mice (average initial weights 36.8 and 26.6 g) were given a single dose of 1200 mg/kg body weight as a suspension in sesame oil by gavage. The treatment gave rise to signs of toxicity such as reduced spontaneous activity, unco-ordinated or atactic gait, prostration, irregular breathing and reduced or abolished righting reflexes. These effects were reversible after 24 hours. Bone marrow from the femurs of 5 mice of each sex was processed 24, 48 and 72 hours after dosing and 1000 polychromatic erythrocytes were evaluated per animal. The ratio of polychromatic to normochromatic erythrocytes was unchanged at all times and there was no increase in the number of micronucleated erythrocytes. Thus in this study, 2,5-dimethoxy-4-chloroaniline had no genotoxic activity (Hoechst, 1993).

7.7   Carcinogenicity

No information available.

7.8   Reproductive toxicity

No information available.

7.9   Effects on the immune system

No information available.

7.10   Neurotoxicity

No information available.

7.11   Other effects

No information available.

8.   Experience in humans

No information available.

9.   Threshold limit values

No information available.

References

Hoechst AG, Pharma Forschung Toxikologie
Akute orale Toxizität von Chloraminohydrochinondimethyläther
(CMEB) an weiblichen Ratten
Unpublished report no. 427/79 (1979)

Hoechst AG, Pharma Forschung Toxikologie
Chloraminohydrochinondimethylether, Prüfung auf Hautreizung
am Kaninchen
Unpublished report no. 86.0293 (1986a)

Hoechst AG, Pharma Forschung Toxikologie
Chloraminohydrochinondimethylether, Prüfung auf Augenreizung
am Kaninchen
Unpublished report no. 86.0294 (1986b)

Hoechst AG, Pharma Research Toxicology
Aminochlorhydrochinondimethylether, Study of the mutagenic
potential in strains of *Salmonella typhimurium* (Ames test) and
Escherichia coli
Unpublished report no. 86.1050 (1986c)

Hoechst AG, Pharma Forschung Toxikologie und Pathologie
Aminochlorhydrochinondimethylether, Prüfung der akuten dermal-
en Toxizität an der Wistar-Ratte
Unpublished report no. 88.0182 (1988a)

Hoechst AG, Pharma Research Toxicology and Pathology
Aminochlorhydrochinondimethylether, Detection of gene muta-
tions in somatic cells in culture, HGPRT-test with V79 cells
Unpublished report no. 88.0811 (1988b)

Hoechst AG
AIDA-Grunddatensatz Benzenamine, 4-chloro-2,5-dimethoxy-(9CI)
(1991a)

Hoechst AG
Written communication to BG Chemie (1991b)

Hoechst AG, Pharma Development Central Toxicology
4-Chlor-2,5-dimethyloxyanilin TF, Micronucleus test in male and
female NMRI mice after oral administration
Unpublished report no. 92.1154 (1993)

HRC (Huntingdon Research Centre Ltd., England)
2,5-Dimethoxy-4-chloraniline (BG catalogue no. 121), fourteen-
day dietary preliminary study in rats
Unpublished report BGH 30/911389 (1993a)
On behalf of BG Chemie

HRC (Huntingdon Research Centre Ltd., England)
2,5-Dimethoxy-4-chloroaniline (BG catalogue No. 121), twenty-
eight day oral toxicity study in rats
Unpublished report BGH 44/920974 (1993b)
On behalf of BG Chemie

MB Research Laboratories Inc., Spinnerstown, Pennsylvania, USA
Test for oral toxicity in rats
Unpublished report, project no. MB 79–3700 (1979)

Schafer, E.W., Bowles, W.A., jr.
Acute oral toxicity and repellency of 933 chemicals to house and
deer mice
Arch. Environ. Contam. Toxicol., 14, 111–129 (1985)

Antimony-V-chloride

1. Summary and assessment

Following intraperitoneal injection in the rat, 88% of pentavalent antimony is excreted in the urine and 1% in the faeces.

Antimony-V-chloride is moderately toxic following acute oral exposure (LD_{50} rat oral 1115 mg/kg body weight). According to an unclear citation in the literature, the 2-hour LC_{50} in the rat is 720 mg/m^3.

A limited subchronic study gave weak indications of partly reversible damage to the lungs, heart, liver, kidneys and thyroid gland as well as accumulation of antimony in the spleen, heart, lymph nodes and thyroid gland, following inhalation exposure of rats to 0.037 mg antimony-V-chloride/l air, 2 hours/day for 4 months.

According to the available genotoxicity studies, antimony-V-chloride is negative in the Salmonella/microsome test in strains TA 98 and TA 100 with and without metabolic activation, and does not increase the sister chromatid exchange rate in mammalian cells in vitro. Of three spot tests with antimony-V-chloride in *Bacillus subtilis*, two were positive and one was negative. Overall, the genotoxic potential of the substance cannot be evaluated based on these studies.

In man, antimony-V-chloride is irritant or corrosive to the skin and mucous membranes of the respiratory tract and can cause pulmonary oedema.

2. Name of substance

2.1	Usual name	Antimony-V-chloride
2.2	IUPAC-name	Antimony-V-chloride

2.3	CAS-No.	7647–18–9
2.4	EINECS-No.	231–601–8

3.   Synonyms, common and trade names

Antimon-V-chlorid
Antimony pentachloride
Antimony (V) chloride
Antimony perchloride
Pentachloroantimony

4.   Structural and molecular formulae

4.1   Structural formula

$$\begin{array}{c} Cl \\ | \\ Cl-Sb{<}^{Cl}_{Cl} \\ | \\ Cl \end{array}$$

4.2   Molecular formula $SbCl_5$

5.   Physical and chemical properties

5.1   Molecular mass, g/mol   299.02

5.2   Melting point, °C   4.0   (Herbst et al., 1985)

5.3   Boiling point, °C   140 (at 1013 hPa)
        (Herbst et al., 1985)

5.4   Vapour pressure, hPa   1 (at 20 °C)
        (Riedel-de Haën, 1988)

5.5   Density, g/cm³   2.34 (at 20 °C)
        (Herbst et al.,1985)

5.6   Solubility in water   Reacts violently with water
        (Riedel-de Haën, 1988)

5.7	Solubility in organic solvents	Soluble in chloroform, carbon tetrachloride (Windholz et al., 1983)
5.8	Solubility in fat	No information available
5.9	pH value	No information available
5.10	Conversion factor	1 ml/m^3(ppm) $\triangleq$ 12.20 mg/m^3 1 mg/m^3 $\triangleq$ 0.08 ml/m^3 (ppm) (at 1013 hPa and 25 °C)

6. Uses

Chlorinating agent in organic chemistry (Herbst et al., 1985)

7. Experimental results

7.1 Toxicokinetics and metabolism

Sprague-Dawley rats were each given a single intraperitoneal injection of 3 μg pentavalent ^{124}antimony. The percentage of the administered radioactivity bound to the blood plasma in dialysable form was 94 and 96% after 45 minutes and 8 hours, respectively. Within 24 hours, 88% of the radioactivity was excreted in the urine and only 1% in the faeces. After a single intravenous injection of 0.1 μg pentavalent ^{125}antimony, ca. 30% had been excreted in the urine and about 1.3% in the bile after 140 minutes (Edel et al., 1993).

In a 4-month inhalation study, 20 rats were exposed to antimony-V-chloride at a concentration of 0.015 mg pentavalent antimony/l air (average value of 30 measurements; equivalent to 0.037 mg antimony-V-chloride/l air; no further details on exposure conditions or particle spectrum of the aerosols). The antimony content of the lung, liver, kidney, spleen, heart, blood, thyroid gland, adrenal glands, lymph nodes and pancreas was determined in groups of 10 animals at the end of the 4-month study and after a 1-month observation period. The highest antimony levels were found immediately after the end of the study in the blood (24.3±5.8 μg antimony/g) and the lungs (19.7±2.87 μg antimony/g wet weight). Lower antimony levels were detected in the other organs (0.73 to 4.8 μg antimony/g). At the end of the observation period, a clear reduction in the antimony content (by about 60%) was detectable only in the blood and the lungs; in the spleen, heart and particularly in the lymph nodes and the thyroid gland, the antimony content was clearly increased at 3.27 to 39.3 μg/g (Tschekunova, 1971).

7.2   Acute and subacute toxicity

The following LD_{50} and LC_{50} values were determined following acute oral or inhalation exposure to antimony-V-chloride. The substance was shown to be moderately toxic on oral administration.

Rat	oral	1115 (961.20–1293.40) mg/kg body weight (Arzamastsev, 1964)
Rat	inhalation	720 mg/m^3 (2 hours; TPI, 1982)
Mouse	inhalation	0.62 mg/l (duration of exposure not specified; Tschekunova and Minkina, 1969)
Guinea pig	oral	900 mg/kg body weight (Arzamastsev, 1964)

Signs of toxicity seen in rats and guinea pigs following oral administration included reduced motility and sedation, with paresis of the hind limbs also seen in the guinea pigs. The animals died mainly during the first 3 days. Histopathological examination of the guinea pigs revealed acute gastroenteritis with necrotic, purulent, inflammatory, catarrhal changes to the gut, congested livers with necrotic foci and degenerative effects on the myocardium and kidneys (no further details; Arzamastsev, 1964).

After exposure to antimony-V-chloride by inhalation, the main effects seen in the mouse were irritant. Post-mortems revealed pulmonary oedema (no further details; Tschekunova and Minkina, 1969).

7.3   Skin and mucous membrane effects

No information available.

7.4   Sensitisation

No information available.

7.5   Subchronic and chronic toxicity

In an inhalation study, 20 rats were exposed to an antimony-V-chloride aerosol at an average concentration of 0.015 mg pentavalent antimony/l air for 2 hours/day for 4 months (equivalent to 0.037 mg antimony-V-chloride/l air; no further details on exposure conditions). A further group of 20 rats served as a control. At the end of the 4-month exposure period, 10 rats/group were killed, with the remaining 10/group observed for a further month. The general condition of the animals was unaffected. Body weight gain during the 4-month exposure was slightly but not significantly reduced. The blood count, the haemoglobin level and the glycogen content in the liver were within the range of control values, although the glycogen content was clearly reduced in a few of the animals (no further details). The relative weights of the liver, kidneys and adrenal glands were unaffected. A slowing of the heart rhythm was detected in 5/19 of the exposed rats after 2 months of the study (ECG in conscious animals) and degenerative changes to the myocardium were found in 4/10 of the exposed animals on histological examination of the heart muscle at the end of the 4-month exposure period. The histological examination also revealed "dystrophy" and pyknosis of the hepatocyte nuclei in the liver, while in the kidney, damage to the epithelium of the kidney tubules with protein deposits and, in some cases, mild interstitial nephritis were evident. The lungs showed congestion, bleeding and oedema and in some places thickening of the alveolar septa. "Dystrophy" was observed in the spleen. In the thyroid gland, the follicles were enlarged and contained dense colloidal material, and the follicular epithelium was enlarged and, in some places, desquamative. In the adrenal glands, oedema and bleeding were seen with enlargement of the cortex and adenomatous changes. The latter were also found in the lymph nodes. No pathological effects were detectable in some of the rats at the end of the 1-month observation period (no further details), although others still showed the effects described but to a lesser degree (Tschekunova and Minkina, 1969).

7.6   Genotoxicity

7.6.1 In vitro

Antimony-V-chloride (99.99% pure) was not mutagenic in a Salmonella microsome preincubation test in strains TA 98 and TA 100 at concentrations of 54, 108, 216 and 432 µg/plate with and without metabolic activation (S9-mix from rat liver, no details of induction). A concentration of 864 µg/plate was cytotoxic (Kuroda et al., 1991; Endo et al., 1991).

In a spot test in *Bacillus subtilis* H17/M45, antimony-V-chloride was mutagenic at 0.03 ml/test plate (equivalent to 70 mg/test plate; Kada et al., 1980; Kanematsu and Kada, 1978; Kanematsu et al., 1980).

Antimony-V-chloride (99.99% pure) was also positive in a further spot-test in *Bacillus subtilis* H17/M45 at concentrations of 65, 130 and 260 µg/test plate (Kuroda et al., 1991; Endo et al., 1991).

In contrast, there were no indications of mutagenic activity in another spot test with antimony-V-chloride in *Bacillus subtilis* H17/M45. The concentration tested was 50 µl of a 0.05 M solution of antimony-V-chloride (equivalent to 748 µg/test plate; no further details; Nishioka, 1975).

Antimony-V-chloride (99.99% pure) was not genotoxic in a sister chromatid exchange test in Chinese hamster V79 cells at concentrations of 8.6, 17.3, 35 and 70 µg/ml, the latter concentration being cytotoxic (Kuroda et al., 1991; Endo et al., 1991).

7.6.2 In vivo

No information available.

7.7   Carcinogenicity

No information available.

7.8   Reproductive toxicity

No information available.

7.9   Effects on the immune system

No information available.

7.10   Neurotoxicity

No information available.

7.11   Other effects

Rats were given a single subcutaneous injection of 3.0, 10.0, 15.0 or 30.0 mg pentavalent antimony (as antimony-V-chloride)/kg body weight (equivalent to 7.4, 24.6, 36.8 and 73.7 mg antimony-V-chloride/kg). This led to a dose-dependent increase in haemoxygenase activity (the enzyme responsible for the degradation of haem to bile pigments) in the liver (maximum 3-fold) 16 hours after dosing. The haemoxygenase activity in the kidney was increased by a maximum of 1.5-fold. The activities of δ-aminolaevulinic acid synthase, ethylmorphine demethylase and aniline hydroxylase and the cytochrome P-450 and microsomal haem contents were not affected (Drummond and Kappas, 1981).

8.   Experience in humans

It has been reported that antimony-V-chloride has an irritant or corrosive effect on the skin and that antimony-V-chloride vapour ("fumes") is severely irritating to the eyes, face and mucous membranes of the respiratory tract (no further details; Grant, 1986).

The exposure of 3 people to antimony-V-chloride following a leak in a reaction vessel caused severe pulmonary oedema and hypoxia and led to the deaths of 2 of the 3 victims, despite treatment (tracheostomy, endotracheal intubation, inhalation of oxygen, corticoid therapy). It is not clear in the report whether the second- and third-degree burns that were evidently suffered in the accident could have been causative (no further details; Cordasco and Stone, 1973; Cordasco, 1974).

9.   Threshold limit values

The MAK value for antimony (Sb) in the Federal Republic of Germany is 0.5 mg total Sb dust/m^3 air (DFG, 1993; TRGS, 1994).
The TLV-TWA value in the USA is also 0.5 mg Sb/m^3 air (ACGIH, 1993–1994).

References

ACGIH (American Conference of Governmental Industrial Hygienists)
Threshold limit values for chemical substances and physical agents and biological exposure indices (1993–1994)

Arzamastsev, E.V.
Experimental substantiation of the permissible concentrations of tri- and pentavalent antimony in water bodies
Hyg. Sanit., 29, 16–21 (1964)

Cordasco, E.M.
Newer concepts in the management of environmental pulmonary edema
Angiology, 25, 590–601 (1974)

Cordasco, E.M., Stone, F.D.
Pulmonary edema of environmental origin
Chest, 64, 182–185 (1973)

DFG (Deutsche Forschungsgemeinschaft)
Maximale Arbeitsplatzkonzentrationen und Biologische Arbeitsstofftoleranzwerte 1993
VCH Verlagsgesellschaft, Weinheim (1993)

Drummond, G.S., Kappas, A.
Potent heme-degrading action of antimony and antimony-containing parasiticidal agents
J. Exp. Med., 153, 245–256 (1981)

Edel, J., Marafante, E., Sabbioni, E., Manzo, L.
Metabolic behaviour of inorganic forms of antimony in the rat
International Conference on Heavy Metals in the Environment, 574–577 (1993)

Endo, G., Kuroda, K., Okamoto, A., Yoo, Y.S., Horiguchi, S.
Genotoxicity of Be, Ga, Sb and As compounds
Mutat. Res., 252 (1), 84–85 (1991)

Grant, W.M.
Toxicology of the eye
3rd ed., pp. 112, 1065–1066
Charles C. Thomas, Publisher, Springfield, Illinois, USA (1986)

Herbst, K.A., Rose, G., Hanusch, K., Schumann, H., Wolf, H.U.
Antimony and antimony compounds
In: Ullmann's encyclopedia of industrial chemistry
5th ed., vol. A3, pp. 55, 67–70, 74–76
VCH Verlagsgesellschaft, Weinheim (1985)

Kada, T., Hirano, K., Shirasu, Y.
Screening of environmental chemical mutagens by the rec-assay system with *Bacillus subtilis*
Chem. Mutagens, Prin. Meth. Detect., 6, 149–173 (1980)

Kanematsu, K., Kada, T.
Mutagenicity of metal compounds
Mutat. Res., 53, 207–208 (1978)

Kanematsu, N., Hara, M., Kada, T.
Rec assay and mutagenicity studies on metal compounds
Mutat. Res., 77, 109–116 (1980)

Kuroda, K., Endo, G., Okamoto, A., Yoo, Y.S., Horiguchi, S.
Genotoxicity of beryllium, gallium and antimony in
short-term assays
Mutat. Res., 264, 163–170 (1991)

Nishioka, H.
Mutagenic activities of metal compounds in bacteria
Mutat. Res., 31, 185–189 (1975)

Riedel-de Haën AG
Safety data sheet Antimon(V)-chlorid (1988)

RTECS (Registry of Toxic Effects of Chemical Substances)
Antimony (V) chloride, RTECS-no. CC5075000
Produced by NIOSH (National Institute for Occupational Safety
and Health) (1990)

TPI (Toxic Parameter Index)
Toxic chemicals under single exposure, p. 22 (1982)
Cited in: RTECS (1990)

TRGS (Technische Regeln für Gefahrstoffe) 900
Grenzwerte in der Luft am Arbeitsplatz
Bundesarbeitsblatt, 6, p. 37 (1994)

Tschekunova, M.P.
Zur Frage des Schicksals der haloiden Verbindungen des Antimons
im Organismus (German translation of the Russian)
Gig. Tr. Prof. Zabol., 15, 31–34 (1971)

Tschekunova, M.P., Minkina, N.A.
Über die Toxizität von Antimon-V-chlorid im chronischen Experi-
ment (German translation of the Russian)
Gig. Tr. Prof. Zabol., 13, 25–29 (1969)

Windholz, M., Budavari, S., Blumetti, R.F., Otterbein, E.S. (eds.)
The Merck index
10th ed., p. 102
Merck & Co., Inc., Rahway, USA (1983)

Antimony-III-chloride

1. Summary and assessment

The absorption of antimony-III-chloride from the gastro-intestinal tract following repeated oral administration to the mouse is low, at 1.7% of the administered daily dose within 24 hours. The half-life for elimination from the whole body following a single inhalation exposure is 140 days in the rat. After intraperitoneal injection in the rat or oral administration in the dairy cow, antimony-III-chloride is mainly excreted in the faeces (33 to 45% in the rat and 82% in the cow), irrespective of the species or the route of administration. Excretion in the urine amounts to 6 to 12% in the rat and 1.1% in the dairy cow. Following intravenous injection in the rat, elimination takes place, in approximately equal proportions in the urine and faeces (22.4 and 24.6%, respectively), with about 10% excreted in the bile (enterohepatic circulation). In dairy cows, excretion takes place almost exclusively in the faeces (82%) after oral administration, while following intravenous injection about 50% is excreted in the urine. Only traces of antimony are found in the milk. An antimony-glutathione complex has been proposed as a urinary metabolite in rats.

Antimony-III-chloride is moderately toxic on acute oral exposure (LD_{50} rat 525 and 675 mg/kg body weight).

In limited studies, signs of anaemia were seen in guinea pigs following oral administration of antimony-III-chloride for 2 and 6 months.

The results of genotoxicity studies on antimony-III-chloride are inconsistent. The substance was negative in the Salmonella/microsome test (in 2 strains), and one negative and two positive results have been obtained in the spot test in *Bacillus subtilis*. A spot test in *Escherichia coli* was also negative. Antimony-III-chloride showed dose-dependent positive effects in the sister chromatid exchange test and was dose-dependently clastogenic in murine bone marrow cells in vivo. Overall, antimony-III-chloride appears to have genotoxic activity, at least in vivo.

On pre- and postnatal oral administration to rats in their drinking water, antimony-III-chloride was toxic to the dams and their offspring, but caused no externally visible malformations. The offspring of mice exposed during pregnancy showed behavioural impairments (effects on motor activity, sensory function, learning ability).

In man, antimony-III-chloride is irritant or corrosive to the skin, eyes and the mucous membranes of the airways. Dermal sensitisation has not been observed.

2   Name of substance

2.1	Usual name	Antimony-III-chloride
2.2	IUPAC-name	Antimony-III-chloride
2.3	CAS-No.	10025–91– 9
2.4	EINECS-No.	233– 047– 2

3.   Synonyms, common and trade names

Antimon-III-chlorid
Antimonous chloride
Antimony butter
Antimony chloride
Antimony trichloride
Butter of antimony
Caustic antimony
Trichlorostibine

4.   Structural and molecular formulae

4.1   Structural formula

$$\begin{array}{c} Cl \\ | \\ Sb \\ Cl \quad \quad Cl \end{array}$$

4.2   Molecular formula $SbCl_3$

5.   Physical and chemical properties

5.1   Molecular mass, g/mol   228.10

5.2   Melting point, °C   73.4   (Weast, 1988/89)

5.3   Boiling point, °C   223   (Herbst et al., 1985)

5.4   Vapour pressure, hPa   1.33 (at 49.2 °C)   (Sax, 1984)
0.16 (at 20 °C)
(Riedel-de Haën, 1989)

5.5   Density, g/cm³   3.140 (at 25 °C)
(Weast,1988/89)

5.6   Solubility in water   601.6 g/100 ml (at 0 °C)
(Weast, 1988/89)
Antimony-III-chloride is only stable in water at high concentrations; on dilution antimony-(III)-oxychloride (SbOCl) precipitates, further dilution leading to the formation of hydrated antimony-III-oxide by hydrolysis
(Herbst et al., 1985)

5.7   Solubility in
organic solvents   Soluble in absolute ethanol, chloroform, benzene, acetone
(Weast, 1988/89)

		soluble in carbon disulfide, dioxane, carbon tetrachloride, ether; insoluble in pyridine, quinoline, organic bases (Windholz et al., 1983)
5.8	Solubility in fat	No information available
5.9	pH value	Acid as a consequence of hydrolysis (Riedel-de Haën, 1989)
5.10	Conversion factor	1 ml/m^3 (ppm) $\triangleq$ 9.31 mg/m^3 1 mg/m^3 $\triangleq$ 0.11 ml/m^3 (ppm) (at 1013 hPa and 25 °C)

6. Uses

Catalyst for polymerisation and chlorination reactions in organic chemistry; used in the manufacture of flame-proofing agents, pigments and other antimony compounds (Herbst et al., 1985).

7. Experimental results

7.1 Toxicokinetics and metabolism

Groups of 6 male Sprague-Dawley rats (200 to 300 g) were given 200, 400 or 800 mg antimony/kg body weight as antimony-III-chloride as a single intraperitoneal injection. Within 96 hours, 10.6 to 12.1% of the amount of antimony administered had been excreted in the urine and 39.1 to 45.2% in the faeces, independent of the dose. The total amount excreted was 49.7 to 56.3 %, with the majority (35.2 to 45.4%) eliminated within the first 24 hours. After a single intravenous injection of 800 µg antimony/kg body weight, the total excreted within the first 96 hours was 22.4% in the urine, and 24.6 % in the faeces (total excretion 47%); about 10% (6 to 15%) of the administered dose was found in the bile within 7 hours, which led the authors to postulate that enterohepatic circulation occurred. Depletion of glutathione in the liver prior to antimony administration resulted in a reduction in faecal excretion and an increase in urinary excretion, while increasing glutathione concentrations in the liver had the opposite effect. In contrast to inorganic arsenic compounds, no methylated antimony compounds were found in the bile or urine by paper chromatography. However, one component was detected with an Rf value similar to that of the antimony-glutathione complex (Bailly et al., 1991).

After a single intraperitoneal injection of 3 µg trivalent [124]Sb/rat, more than 95% of the activity was bound to the erythrocytes in dialysable form 2 hours after administration. During the first day after injection, 33% of the activity was excreted in the faeces and 6% in the urine (Edel et al., 1993).

Mice (BALB/c) were given 18.5 MBq [125]antimony-III-chloride/kg feed and 37 MBq H_3[75]AsO_4/kg feed from days 1 to 7 of pregnancy. Radioactivity was measured in the gut and other tissues on days 3, 5 and 7 of pregnancy (number of animals used not specified). The maximum values were reached on day 5 of pregnancy. On this day, the radioactive antimony detected in the gut, at a maximum of 36% of the daily dose, was clearly higher than that in the other tissues, at 1.92%. The daily absorption rate from the gut was calculated to be 1.7% of the daily dose of antimony. In the lungs, bones and ovaries and uterus, the antimony concentrations were ca. 0.085, 0.14 and 0.1 to 0.2% of the daily dose/g organ (Gerber et al., 1982).

Pregnant mice were given 18.5 MBq [125]antimony-III-chloride/kg feed and 37 MBq H_3[75]AsO_4/kg feed for the entire duration of pregnancy and up until day 15 of lactation. The radioactivity was then monitored for 150 days in the dams and 50 days in their offspring, with measurements made after 10, 20, 50 and 150 study days (numbers of animals used not specified). The whole-body elimination half-life for antimony was 96±48 days in the dams and ca. 10 days in the offspring. The radioactivity in the organs of the dams was often present at the limits of detection, measurable antimony activities being found in the thyroid gland (no further details). In the offspring, the highest antimony levels (radioactivity/g tissue) were found in the bones, muscle, spleen, heart, kidney and lung (no further details). After 50 days, radioactivity was detectable only in the muscle, skin and spleen of the offspring. When litters from dams exposed to radioactivity were exchanged with unexposed litters following birth, levels of radioactive antimony corresponding to 80% of those in the offspring exposed in utero were reached within a short time (no further details; Gerber et al., 1982).

BALB/c mice were given a single intraperitoneal injection of 20 kBq ^{125}antimony-III-chloride and 42 kBq $H_3{}^{75}AsO_4$/mouse on day 12 of pregnancy. The radioactivity in the foetus and the placenta and in the kidney, gut, spleen, liver, lung, skin, bones, muscle tissue, brain, blood, uterus and ovaries of the dams was measured 30 minutes, 3, 6 and 24 hours and 2, 3, 4 and 7 days after dosing (numbers of animals used not specified). Whole-body measurements gave elimination half-lives for antimony of ca. 6 hours for about 95% of the administered dose and 2.4±0.3 days for the remaining 5%. The radioactive antimony measured in the bones of the dams 3 hours after administration was equivalent to ca. 4% of the administered dose/g tissue, declining to less than 1% 7 days after administration, while that in the gut declined from 30 to 0.1%. In the uterus and ovaries, the radioactive antimony declined only slightly to 1 to 2% of the administered dose/g tissue over a period of 7 days following administration compared with 2 to 3%, 6 hours after dosing. The other organs (lung, brain, kidney, spleen, liver, blood, skin, muscle) and the placenta and foetuses contained maximum levels of 0.01 to 0.7% of the administered radioactive antimony/g tissue 6 to 24 hours after dosing (Gerber et al., 1982).

3 lactating dairy cows were given single oral doses of 2.00, 2.72 and 2.84 mCi (7.4×10^7 Bq, 10.1×10^7 Bq and 10.5×10^7 Bq, respectively) ^{124}antimony-III-chloride/animal, while a further cow was given a single intravenous injection of 0.234 mCi (0.87×10^7 Bq) ^{124}antimony-III-chloride. The observation periods were 102 days (2 animals) and 140 days (1 animal) after oral administration and 70 - days after intravenous injection. After oral administration, ca. 82% of the administered dose was excreted via the faeces and ca. 1.1% in the urine within 140 days, while after intravenous injection, ca. 2.4% was excreted in the faeces and ca. 51% in the urine within 70 days. Half-lives for elimination in the faeces were ca. 29.1 days

after oral administration and 30.5 days after intravenous injection. In the urine, the half-lives were 4.64 and 24.0 days after oral and intravenous administration, respectively. Ca. 0.008 and 0.05% of the oral and intravenous doses, respectively, were secreted in the milk. The elimination half-lives in the milk were 28.0 days after oral administration and 9.2 days after intravenous injection. The distribution of radioactivity was determined in the tissues after 102 days (oral administration) and 70 days (intravenous injection) and is shown in Table 1.

Table 1. Distribution of radioactivity in dairy cows after oral and intravenous administration of ^{124}antimony trichloride

Tissue	% of activity determined in the body/organ		% of administered dose/kg tissue x 10^{-4}	
	oral	intravenous*	oral	intravenous
Skin	43.3	2.67 / 6.67	1.65	53
Femur	30.1		1.65	
Muscle	10.1	2.73 / 6.98	0.16	23.2
Liver	7.32	25.2 / 69.4	2.88	5160
Small intestine	1.19	0.16 / 0.42	0.62	44
Spleen	1.05	2.52 / 6.44	3.99	4990
Lung	1.07	0.40 / 1.03	0.56	109
Heart	0.08	60.8	0.10	42000*

*   A high proportion of the antimony dose was deposited in the heart, the values given were calculated with and without this amount, which was linked with the site of injection

The radioactivity determined in the other tissues was on average 0.28 and 104.5x10^{-4}% of the orally or intravenously administered dose/kg tissue (van Bruwaene et al., 1982).
In an acute inhalation study, groups of 20 female rats were exposed to 20 mCi (7.4x10^8 Bq) ^{124}antimony-III-chloride as an aqueous aerosol for 30 minutes. The median aerodynamic particle diameter was between 0.02 and 0.04 μm. The observation period was 140 - days. Whole-body radioactivity was determined daily. In addition, the radioactivity in the individual tissues was determined after 1, 2, 4, 8, 15, 22 and 140 days. The elimination half-life from the lungs was $\geq$ 90 days and from the whole body was 140 days. The radioactivity was predominantly found in the lungs, in whole blood, in the spleen and in the heart (Djuric et al., 1962).

In a further inhalation study, guinea pigs were exposed to antimony-III-chloride dust at 100 mg/m^3 for 12 minutes. The median aerodynamic particle diameter was 1 μm. Dust retention was monitored for up to 18 hours after exposure. Antimony-III-chloride had been almost completely cleared from the lungs 6 hours after exposure (no further details; Palm, 1954).

7.2   Acute and subacute toxicity

The following LD$_{50}$ values have been determined following acute exposure to antimony-III-chloride. The substance is moderately toxic on oral administration.

Mouse	i.p.	13 mg/kg body weight (TPI, 1982)
Mouse	oral	700 mg/kg body weight (Gurnani et al., 1992)
Rat	oral	525 mg/kg body weight (Marhold, 1977)
Rat	oral	675 (557.8 to 816.75) mg/kg body weight (Arzamastsev, 1964)
Guinea pig	oral	574 mg/kg body weight (Arzamastsev, 1964)

Signs of toxicity observed in the rats and guinea pigs included sedation and reduced motility, with paresis of the hind limbs also observed in the guinea pigs. The animals died mainly during the first 3 days after administration. Histopathological examination of the guinea pigs following oral administration revealed acute gastroenteritis with necrotic, purulent inflammatory, catarrhal changes in the gut, congestion of the liver with necrotic foci and degenerative changes in the myocardium and kidneys (Arzamastsev, 1964).

7.3   Skin and mucous membrane effects

No information available.

7.4   Sensitisation

No information available.

7.5   Subchronic and chronic toxicity

The oral administration of 12 or 20 mg trivalent antimony/kg body weight/day (as antimony-III-chloride, equivalent to 22.4 and 37.4 mg antimony-III-chloride/kg body weight/day, respectively) to guinea pigs by gavage for 2 months led to decreased body weight gain. The haemoglobin content was significantly reduced ($p < 0.05$) at the end of the second month of the study, while the number of erythrocytes tended to be reduced. The number of reticulocytes was significantly increased ($p < 0.05$). There were no effects on the differential blood count. Analysis of the serum protein fractions revealed a gradual increase in the gamma-globulin fraction, which in the high dose group had increased by more than 2-fold at the end of the study. A tendency towards an increase in the cholesterol level in the serum could only be detected at the end of the first month of the study. The amino acid content of the urine remained unaffected. A reduction in the level of free SH-groups in the serum occurred dose-dependently from day 15 of the study and reached a maximum of 30% at the end of the study. The mitotic index in the bone marrow was dose-dependently reduced by a maximum of ca. 27% at the end of the study. The activities of alanine- and aspartate aminotransferase (ALT and AST) were increased in the serum. Electrocardiograms gave indications of myocardial effects. No histological examinations were carried out (no further details; Arzamastsev, 1964).

Groups of 17 guinea pigs were given antimony-III-chloride at doses of 0.0025, 0.025, 0.25 and 2.5 mg Sb^{3+}/kg body weight/day orally for 6 months (equivalent to 0.0047, 0.047, 0.47 and 4.7 mg antimony-III-chloride/kg body weight/day). The level of free SH-groups in the serum was dose-dependently reduced from 0.047 mg/kg body weight, the reduction being statistically significant ($p < 0.05$) at the top dose level (4.7 mg/kg body weight) from the first month of the study. In the animals in the two higher dose groups (0.47 and 4.7 mg/kg body weight), there was an increase in the reticulocyte count and a fall in the erythrocyte count and the haemoglobin content. At the end of the study, the high dose animals showed an increase in the gamma-globulin fraction with a simultaneous decrease in the total protein content of the serum. At the two higher dose levels there was an increase in the threshold of stimulation of the action potential in electroencephalograms. Overall, 0.047 mg/kg was given as the "threshold dose" and 0.0047 mg/kg as the no effect level (Arzamastsev, 1964).

7.6   Genotoxicity

7.6.1 In vitro

Antimony-III-chloride was not genotoxic in the Salmonella/microsome test in strains TA 98 and TA 100 at concentrations ranging from 625 to 5000 μg/plate, in the preincubation test with and without metabolic activation (S9-mix from rat liver, no details of induction; Kuroda et al., 1991; Endo et al., 1991).

A concentration of 0.01 M antimony-III-chloride (equivalent to 114 μg/plate at a volume of 0.05 ml/plate) was mutagenic in the spot test in *Bacillus subtilis* H17/M45. No further concentrations were tested (Kada et al., 1980; Kanematsu and Kada, 1978; Kanematsu et al., 1980).

In a further spot test in *Bacillus subtilis* H17/M45, 98% pure antimony-III-chloride was also positive at doses of 6.3, 12.5 and 23 μg/well (Kuroda et al., 1991; Endo et al., 1991).

In contrast, there were no indications of mutagenic activity in another spot test with antimony-III-chloride in *Bacillus subtilis* H17/M45. The concentration tested was 0.05 M antimony-III-chloride, equivalent to 570 μg/plate at a volume of 0.05 ml/plate (Nishioka, 1975).

Spot tests in *Escherichia coli* (B/r WP2, WP2) and in *Salmonella typhimurium* (TA 98, TA 100, TA 1535, TA 1537 and TA 1538) were negative. It is not possible to tell whether doses in the cytotoxic range were tested, due to a lack of data on concentration (Kanematsu et al., 1980).

In a sister chromatid exchange test in Chinese hamster V79 cells, 98% pure antimony-III-chloride was negative at concentrations of 1.3 and 2.5 μg/ml, but positive at 5 and 10 μg/ml. A concentration of 20 μg/ml was toxic (Kuroda et al., 1991; Endo et al., 1991).

7.6.2 In vivo

The ability of antimony-III-chloride to damage chromosomes in murine bone marrow cells was studied in vivo. Groups of 5 female

Swiss mice (25 to 30 g) were given single doses of 0, 70, 140 and 233.33 mg antimony-III-chloride/kg body weight by gavage. The chromosomes from the bone marrow of the femur were studied (100 metaphases/mouse) 6, 12, 18 and 24 hours later. 200 cells/mouse were evaluated to monitor cell division. There were dose-dependent increases in chromosome aberrations and gaps ($p \leq 0.001$), and in chromosome breaks/cell ($p \leq 0.01$ to 0.001) at all time points studied. The rate of cell division was not significantly changed. Antimony-III-chloride was therefore clastogenic in this study, even at doses of $^1/_{10}$ of the LD_{50} (700 mg/kg; Gurnani et al., 1992).

The administration of a single oral dose of 500, 1000 or 1500 mg antimony-III-chloride/kg body weight to Swiss mice (5 to 14 animals/dose) caused a dose-dependent inhibition of ^{3}H-thymidine incorporation in isolated spleen cells during DNA repair after UV irradiation and during replicative synthesis 24 hours after administration. There was also a dose-dependent increase in fragmentation of the DNA in the nucleoli which, at the highest dose, could not be increased by gamma-irradiation; "rejoining" of the DNA strand breaks was delayed following gamma-irradiation. After 72 hours, all of the values were again within the range of those of the controls (Ashry et al., 1988).

7.7 Carcinogenicity

No information available.

7.8 Reproductive toxicity

Female albino rats (10 to 30 animals/dose) were given 0.1 or 1.0 mg antimony-III-chloride/100 ml drinking water (equivalent to 0.12 and 1.2 mg/kg body weight/day, respectively, at an assumed drinking water intake of 120 ml/kg body weight/day) from day 1 of pregnancy to postnatal day 22. In a second experiment, the same doses were administered from the day of birth until postnatal day 22. The offspring of the dams in both studies were given 0.1 and 1.0 mg antimony-III-chloride/100 ml drinking water, respectively, from day 22 to day 60 of life. The control groups in each case consisted of 10 to 30 dams. In the dams that were given antimony-III-chloride from day 1 of pregnancy onwards (experiment A), body weight gain was dose-dependently significantly reduced (p<0.05) by a maximum of 10% on day 20 of pregnancy, while in the offspring in the high dose group (1.2 mg/kg body weight) it was reduced by a maximum of 47% from day 10 of life onwards. Body weight gain in the dams given antimony-III-chloride from the day of birth onwards (experiment B) was within the range of control values, as was that in their offspring. The duration of pregnancy, number of offspring/litter and systolic arterial blood pressure in the dams and their offspring were unimpaired. No external malformations were evident after birth. Dose-dependent, significant reductions (p<0.05) in the hypertonic effect of 1-noradrenaline (0.1 to 5.0 µg/kg body weight i.v.), by a maximum of ca. 46 and 26%, respectively, were found on day 60 of life in the offspring of dams in experiments A (peri- and postnatal exposure) and B (postnatal exposure). Following carotid occlusion (40 seconds) on day 60 of life, significant inhibition (15%, p<0.05) of induced hypertonia was observed only in the offspring postnatally exposed to antimony-III-chloride at the high dose (1.2 mg/kg body weight) and not in those exposed perinatally and postnatally. The hypotonic effect of 1-isoprenaline (0.01 to 1.0 µg/kg body weight i.v.) was dose-dependently significantly inhibited (p<0.05) in offspring of dams in experiment A on day 60 of life at both dose levels by a maximum of ca. 31% while that of acetyl choline (0.01 to 1.0 µg/kg body weight i.v.) was only inhibited at the high dose (1.2 mg/kg body weight/day) by a maximum of 14%. In the offspring of dams in experiment B, these effects occurred in the high dose group as early as day 30 of life, the inhibition of the responses to 1-isoprenaline and acetyl choline reaching maxima of 21% and 44%, respectively (Angrisani et al., 1988; Marmo et al., 1987; Rossi et al., 1987).

Mice were treated with 0.008 ng trivalent antimony/animal/day (equivalent to 0.75 ng antimony-III-chloride/kg body weight/day at an assumed body weight of 20 g) on 3 consecutive days during gestation and also 8 times later in pregnancy (no further details). The behaviour of 24 of the offspring was studied from days 1 to 16 of life. Significant behavioural impairment (increased activity, shorter latency of "passive avoidance", reduced learning ability in the "open-field test", the rotarod test, the "maze test" and in various other behavioural tests) was observed in the offspring of antimony-III-chloride-treated dams. There were no significant differences in birth weight between the offspring of treated mothers compared with those of controls. While the weight of the female offspring was slightly increased on day 21 of life, that of the male offspring was within the range of the control values (Tsujii and Hoshishima, 1979).

The LD_{50} values for antimony trichloride in chicken embryos (white leghorn) were 0.10 and 0.50 mg/egg after administration into the yolk sac of 4- and 8-day-old embryos, respectively, and 0.018 mg/egg after administration into the chorio-allantoic membrane of 8-day-old embryos. The observation period covered 18 days of incubation, and 6 to 10 eggs were used per dose (Ridgway and Karnofsky, 1952).

7.9   Effects on the immune system

No information available.

7.10 Neurotoxicity

No information available.

7.11 Other effects

The administration to rats of a single subcutaneous injection of 3.0, 10.0, 15.0 or 30.0 mg trivalent antimony/kg body weight (equivalent to 5.6, 18.7, 28.1 and 56.1 mg antimony-III-chloride/kg) led to a maximum 10-fold dose-dependent increase in haemoxygenase activity (the enzyme responsible for the oxidative degradation of haem to bile pigments) in the liver and kidney, 16 hours after dosing. The microsomal haem content was reduced by ca. 15%. The activity of δ-aminolaevulinic acid synthase in the liver and kidney was increased by a maximum of ca. 2- and 3-fold, respectively, 16 and 24 hours, respectively, after administration, while the cytochrome P-450 contents in the liver and kidney were reduced by ca. 40% and 30%, respectively, 16 hours after administration. The activities of aniline hydroxylase and ethylmorphine demethylase were also reduced by about 40% in the liver 16 hours after administration, while the cytochrome b_5 content remained unaffected. The authors concluded from these results that antimony-III-chloride interferes with haem metabolism in a similar manner to cobalt (Drummond and Kappas, 1981).

8.   Experience in humans

Antimony-III-chloride has an irritating or corrosive effect on the skin. Antimony-III-chloride vapour ("fumes") is severely irritating to the eyes, face and mucous membranes of the respiratory tract. Severe burns of the cornea can occur on exposure to high concentrations of the vapour (no further details; Grant, 1986).

The minimum lethal dose (MLD) of antimony-III-chloride for humans was given as 131 mg/m^3 air (equivalent to 70.1 mg Sb^{3+}/m^3 air) for a 4-hour inhalation exposure (no further details; Venugopal and Luckey, 1978).

In a refinery, antimony-III-chloride was passed through pipes in an enclosed system as a 98% solution in anhydrous hydrochloric acid at ca. 100 °C and 6.9 to 13.8 bar. During maintenance operations 7 workers, who were not wearing protective clothing or breathing apparatus, were briefly exposed to antimony-III-chloride/HCl vapour which escaped from leaking pumps. Subsequent measurements established possible maximum concentrations of 146 mg HCl and 73 mg antimony/m^3 air during the accident. First and second degree burns of the face, neck and upper-body cleared up within 3 to 10 days following treatment (rinsing, magnesium oxide cream). Signs of toxicity that occurred after exposure (irritation of the upper respiratory tract, nausea, vomiting, anorexia, headaches, abdominal pain, loss of appetite) also cleared up completely after 3 to 10 days. Blood counts and chest X-rays were normal. Maximum antimony concentrations of 2.3 and 5.1 µg/ml were detected in the urine of 2 of the exposed workers 1 and 2 days after exposure, respectively; no antimony was detectable (detection limit unspecified) in the first subject after a further 2 days, while the concentration in the urine of the second subject was about 0.01 µg/ml after a further 6 days. In a control experiment in the same plant following a 10-week disruption-free working period, only 2 of 19 workers had maximum antimony concentrations in the urine of 0.20 and 0.31 µg/ml, respectively. No antimony could be detected in the remaining subjects. It was unclear whether the symptoms observed were due to antimony-III-chloride and/or hydrochloric acid exposure or whether they were connected with the burns suffered (Taylor, 1966).

The allergic effect of metal alloys and their components was studied in patch tests in 17 dentists on the permanent staff of a dental clinic. The effects were evaluated after 30 minutes and 24 hours in accordance with the international standard of the ICDRG (International Contact Dermatitis Research Group). Antimony-III-chloride (2% in yellow vaseline) was not sensitising to the skin of the forearm (Shigeto et al., 1989).

A further skin sensitisation study with metal alloys was carried out in 95 subjects (75 men and 20 women; 76 were dental students, 12 were staff at the clinic and 7 were clinic patients). The samples were applied to the skin of the back under a patch for 48 hours and the reactions were read after 48 and 72 hours and, in one case, after 5 to 7 days. Antimony-III-chloride (1% in yellow vaseline) caused erythema in two cases and erythema and oedema in one, but it did not cause any allergic reactions. Thus in this study, antimony-III-chloride did not induce skin sensitisation (Namikoshi et al., 1990).

9.   Threshold limit values

The MAK value for antimony (Sb) in the Federal Republic of Germany is 0.5 mg total Sb dust/m^3 air (DFG, 1993; TRGS, 1994).
The TLV value in the USA is also 0.5 mg Sb/m^3 air (ACGIH, 1993–1994).
The recommended limit value for antimony ions in drinking water in the former USSR was 0.05 mg/l (Arzamastsev, 1964).

References

ACGIH (American Conference of Governmental Industrial Hygienists)
Threshold limit values for chemical substances and physical agents and toxicological exposure indices (1993–1994)

Angrisani, M., Lampa, E., Lisa, M., Matera, C., Marrazzo, R., Scafuro, M.A., Rossi, F., Marmo, E.
Vasomotor reactivity and postnatal exposure to antimony trichloride
Curr. Ther. Res., 43, 153–159 (1988)

Arzamastsev, E.V.
Experimental substantiation of the permissible concentrations of tri- and pentavalent antimony in water bodies
Hyg. Sanit., 29, 16–21 (1964)

Ashry, H.A., Topaloglou, A., Abdelmoati, E., Teherani, D.K., Altmann, H.
Genotoxische Wirkungen von Antimonionen
Report project no. 600016
Österreichisches Forschungszentrum Seibersdorf (1988)
NTIS PB 89–222210

Bailly, R., Lauwerys, R., Buchet, J.P., Mahieu, P., Konings, J.
Experimental and human studies on antimony metabolism: their relevance for the biological monitoring of workers exposed to inorganic antimony
Br. J. Ind. Med., 48, 93–97 (1991)

Bruwaene, van, R., Gerber, G.B., Kirchmann, R., Colard, J.
Metabolism of antimony-124 in lactating dairy cows
Health Phys., 43, 733–738 (1982)

DFG (Deutsche Forschungsgemeinschaft)
Maximale Arbeitsplatzkonzentrationen und Biologische Arbeitsstofftoleranzwerte 1993
VCH Verlagsgesellschaft, Weinheim (1993)

Djuric, D., Thomas, R.G., Lie, R.
The distribution and excretion of trivalent antimony in the rat following inhalation
Int. Arch. Gewerbepathol. Gewerbehyg., 19, 529–545 (1962)

Drummond, G.S., Kappas, A.
Potent heme-degrading action of antimony and antimony-containing parasiticidal agents
J. Exp. Med., 153, 245–256 (1981)

Edel, J., Marafante, E., Sabbioni, E., Manzo, L.
Metabolic behaviour of inorganic forms of antimony in the rat
International Conference on Heavy Metals in the Environment, 574–577 (1993)

Endo, G., Kuroda, K., Okamoto, A., Yoo, Y.S., Horiguchi, S.
Genotoxicity of Be, Ga, Sb and As compounds
Mutat. Res., 252 (1), 84–85 (1991)

Franks, W.R.
Differential susceptibility of animals to dust
A.M.A. Arch. Ind. Health, pp. 288–296 (1955)

Gerber, G.B., Maes, J., Eykens, B.
Transfer of antimony and arsenic to the developing organism
Arch. Toxicol., 49, 159–168 (1982)

Grant, W.M.
Toxicology of the eye
3rd ed., pp. 112, 1065–1066, 1068
Charles C. Thomas, Publisher, Springfield, Illinois, USA (1986)

Gurnani, N., Sharma, A., Talukder, G.
Cytotoxic effects of antimony trichloride on mice in vivo
Cytobios, 70, 131–136 (1992)

Herbst, K.A., Rose, G., Hanusch, K., Schumann, H., Wolf, H.U.
Antimony and antimony compounds
In: Ullmann's encyclopedia of industrial chemistry
5th ed., vol. A3, pp. 55, 57–70, 74–76
VCH Verlagsgesellschaft, Weinheim (1985)

Kada, T., Hirano, K., Shirasu, Y.
Screening of environmental chemical mutagens by the rec-assay
system with *Bacillus subtilis*
Chem. Mutagens, Prin. Meth. Detect., 6, 149–173 (1980)

Kanematsu, K., Kada, T.
Mutagenicity of metal compounds
Mutat. Res., 53, 207–208 (1978)

Kanematsu, N., Hara, M., Kada, T.
Rec assay and mutagenicity studies on metal compounds
Mutat. Res., 77, 109–116 (1980)

Kuroda, K., Endo, G., Okamoto, A., Yoo, Y.S., Horiguchi, S.
Genotoxicity of beryllium, gallium and antimony in short-term
assays
Mutat. Res., 264, 163–170 (1991)

Marhold, J.V.
Personal communication (1977)
Cited in: RTECS (1990)

Marmo, E., Matera, M.G., Acampora, R., Vacca, C., De Santis, D.,
Maione, S., Susanna, V., Chieppa, S., Guarino, V., Servodio, R.,
Cuparencu, B., Rossi, F.
Prenatal and postnatal metal exposure: effect on vasomotor reactiv-
ity development of pups
Curr. Ther. Res., 42, 823–838 (1987)

Namikoshi, T., Yoshimatsu, T., Suga, K., Fujii, H.
The prevalence of sensitivity to constituents of dental alloys
J. Oral Rehabil., 17, 377–381 (1990)

Nishioka, H.
Mutagenic activities of metal compounds in bacteria
Mutat. Res., 31, 185–189 (1975)

Palm, P.E.
Thesis
University of Pittsburgh (1954)
Cited in: Franks (1955)

Ridgway, L.P., Karnofsky, D.A.
The effects of metals on the chick embryo: toxicity and production
of abnormalities in development
Ann. N Y Acad. Sci., 55, 203–215 (1952)

Riedel-de Haën AG
DIN-safety data sheet Antimon(III)-chlorid Stücke, rein (1989)

Rossi, F., Acampora, R., Vacca, C., Maione, S., Matera, M.G.,
Servodio, R., Marmo, E.
Prenatal and postnatal antimony exposure in rats: effect on vasomo-
tor reactivity development of pups
Teratogenesis Carcinog. Mutagen., 7, 491–496 (1987)

RTECS (Registry of Toxic Effects of Chemical Substances)
Antimony trichloride, RTECS-no. CC4900000
Produced by NIOSH (National Institute for Occupational Safety
and Health) (1990)

Sax, N.I.
Dangerous properties of industrial materials
6th ed., p. 300
Van Nostrand Reinhold, New York, NY (1984)

Shigeto, N., Numata, T., Abe, M., Murata, H., Hamada, T., Kioka, M.
Metal hypersensitivity in dentists: a patch test study
Hiroshima J. Med. Sci., 38 (4), 187–189 (1989)

Taylor, P.J.
Acute intoxication from antimony trichloride
Br. J. Ind. Med., 23, 318–321 (1966)

TPI (Toxic Parameter Index)
Toxic chemicals under single exposure, p. 23 (1982)
Cited in: RTECS (1990)

TRGS (Technische Regeln für Gefahrstoffe) 900
Grenzwerte in der Luft am Arbeitsplatz
Bundesarbeitsblatt, 6, p. 37 (1994)

Tsujii, H., Hoshishima, K.
The effect of the administration of trace amounts of metals to pregnant mice upon the behavior and learning of their offspring
Shinshu Daigaku Nogakubu Kiyo, 16, 13–28 (1979)
Venugopal, B., Luckey, T.D.
Metal toxicity in mammals
Vol. 2 Chemical toxicity of metals and metalloids, pp. 213–215
Plenum Press, New York (1978)

Weast, R.C. (ed.)
CRC Handbook of chemistry and physics
69th ed., p. B-72
CRC-Press, Boca Raton, Florida (1988/89)

Windholz, M., Budavari, S., Blumetti, R.F., Otterbein, E.S. (eds.)
The Merck index
10th ed., pp. 103–104
Merck & Co., Inc., Rahway, USA (1983)

Compound chemical index vol. 1-11

CAS-number index vol. 1-11

74-31-7	*7*, 2	97-74-5	*10*, 84
75-02-5	*2*, 1	98-07-7	*4*, 39
75-25-2	*3*, 2	98-08-8	*7*, 27
75-78-5	*9*, 2	98-13-5	*9*, 69
75-79-6	*9*, 14	98-15-7	*3*, 41
77-73-6	*1*, 23; *10*, 3	98-73-7	*5*, 30; *8*, 54
78-79-5	*2*, 14	98-87-3	*4*, 58
78-83-1	*1*, 44	98-88-4	*4*, 68
79-07-2	*1*, 59	99-09-2	*2*, 88
79-10-7	*2*, 41	99-51-4	*9*, 77
79-11-8	*6*, 3	99-52-5	*2*, 101
79-20-9	*10*, 32	99-59-2	*1*, 155
80-17-1	*6*, 33	100-02-7	*6*, 54
81-20-9	*9*, 24	100-21-0	*3*, 47
83-41-0	*9*, 32	100-44-7	*4*, 81
84-65-1	*11*, 5	101-54-2	*5*, 46
87-02-5	*10*, 67	101-67-7	*5*, 65
88-10-8	*3*, 19	101-96-2	*1*, 166
88-16-4	*3*, 27	102-50-1	*1*, 173
89-58-7	*9*, 39	102-71-6	*4*, 126
89-63-4	*1*, 75	103-71-9	*4*, 152
89-87-2	*9*, 46	104-12-1	*4*, 164
90-51-7	*10*, 75	104-76-7	*1*, 182
91-15-6	*2*, 76; *11*, 72	104-91-6	*2*, 108
91-29-2	*11*, 99	105-39-5	*5*, 71
95-52-3	*8*, 1	105-58-8	*7*, 35
95-68-1	*8*, 10	106-20-7	*10*, 121
96-34-4	*3*, 33; *9*, 56	106-47-8	*10*, 132
96-45-7	*1*, 83	106-75-2	*8*, 74
96-48-0	*1*, 134	107-19-7	*2*, 121
97-00-7	*5*, 3	107-25-5	*5*, 83
97-36-9	*8*, 42	108-80-5	*7*, 49
97-39-2	*6*, 41	109-59-1	*3*, 61

Springer and the environment

At Springer we firmly believe that an international science publisher has a special obligation to the environment, and our corporate policies consistently reflect this conviction.
We also expect our business partners – paper mills, printers, packaging manufacturers, etc. – to commit themselves to using materials and production processes that do not harm the environment. The paper in this book is made from low- or no-chlorine pulp and is acid free, in conformance with international standards for paper permanency.

If you have any concerns about our products,
you can contact us on
ProductSafety@springernature.com

In case Publisher is established outside the EU,
the EU authorized representative is:
Springer Nature Customer Service Center GmbH
Europaplatz 3, 69115 Heidelberg, Germany

Printed by Libri Plureos GmbH
in Hamburg, Germany